BACKGROUND PAPER
Y0-CAN-439
Biomedical Ethics in U.S. Public Policy
OFFICE OF TECHNOLOGY ASSESSMENT
CONGRESS OF THE UNITED STATES

Recommended Citation:
U.S. Congress, Office of Technology Assessment, *Biomedical Ethics in U.S. Public Policy — Background Paper*, OTA-BP-BBS-105 (Washington, DC: U.S. Government Printing Office, June 1993).

For sale by the U.S. Government Printing Office
Superintendent of Documents, Mail Stop: SSOP, Washington, DC 20402-9328
ISBN 0-16-041775-9

Foreword

Over the past two decades, a desire for mechanisms to articulate common values and foster consensus about biomedical advances in the face of cultural and religious heterogeneity resulted in the creation of Federal bioethics commissions. In hindsight, clearly some of these efforts have had lasting, measurable impacts. For over a decade, though, no such initiative has been functionally operational.

Recently, however, Congress has renewed its interest in a bioethics commission—signaling, in part, the increasing importance of medical and biological technologies in daily life. In September 1992, Senator Mark O. Hatfield, Ranking Minority, Committee on Appropriations; Senator Edward M. Kennedy, Chairman, Committee on Labor and Human Resources; and Senator Dennis DeConcini, Chairman, Subcommittee on Patents, Copyrights, and Trademarks, Committee on the Judiciary, asked the Office of Technology Assessment (OTA) to examine past, broad-based bioethics entities in the context of the question: If Congress decides to create a new Federal bioethics body, what does past experience reveal about which particular factors promote success and which should be avoided?

OTA prepared *Biomedical Ethics in U.S. Public Policy* with the assistance of workshop participants, contractors, and reviewers selected for their expertise and diverse points of view. Additionally, scores of individuals cooperated with OTA staff through interviews or by providing written material. OTA gratefully acknowledges the contribution of each of these individuals. As with all OTA Reports, however, responsibility for the content is OTA's alone.

The Report reviews the history of four Federal bioethics initiatives: the National Commission for the Protection of Human Subjects of Biomedical and Behavioral Research, the Ethics Advisory Board, the President's Commission for the Study of Ethical Problems in Medicine and Biomedical and Behavioral Research, and the Biomedical Ethics Advisory Committee. Today, as Congress considers whether to create a new Federal bioethics body, it can be guided by considering the strengths and weaknesses of these efforts. We believe that lessons from the past will prove instructive for the future. As the frontiers of biomedical research and technology continue to advance, it will become increasingly important for policymakers and the public to understand the ethical implications of such innovation.

Roger C. Herdman, Director

Workshop Participants

David Blumenthal, *Workshop Chair*
Massachusetts General Hospital
Boston, MA

Adrienne Asch
Boston University School of Social Work
Boston, MA

Alexander M. Capron
University of Southern
California Law Center
Los Angeles, CA

Robert M. Cook-Deegan
Institute of Medicine
Washington, DC

Ruth R. Faden
Johns Hopkins University School of
Hygiene and Public Health
Baltimore, MD

John C. Fletcher
University of Virginia
Charlottesville, VA

Jonathan Glover
Oxford University
Oxford, United Kingdom

Patricia A. King
Georgetown University Law Center
Washington, DC

Bartha Maria Knoppers
University of Montreal Law Faculty
Montreal, Canada

Charles R. McCarthy
Kennedy Institute of Ethics
Washington, DC

Ellen H. Moskowitz
The Hastings Center
Briarcliff Manor, NY

Gail J. Povar
George Washington University
Medical Center
Washington, DC

Donald W. Seldin
University of Texas Southwestern
Medical School
Dallas, TX

David Shapiro
Nuffield Council on Bioethics
London, United Kingdom

Michael S. Yesley
Los Alamos National Laboratory
Los Alamos, NM

NOTE: OTA appreciates and is grateful for the valuable assistance provided by the workshop participants. The workshop participants do not, however, necessarily approve, disapprove, or endorse this report. OTA assumes full responsibility for the report and the accuracy of its contents.

Project Staff

Robyn Y. Nishimi
Project Director

Ellen L. Goode
Research Analyst

Michael Gough
Program Manager
Biological and Behavioral Sciences

Cecile Parker
Office Administrator

Linda Rayford-Journiette
PC Specialist

Jene Lewis
Administrative Secretary

CONTRACTORS

Kathi E. Hanna
Churchton, MD

Constance M. Pechura
Editor
Bethesda, MD

Daniel J. Wikler
University of Wisconsin
Madison, WI

Contents

Tables

APPENDIXES

Introduction 1

Ethics and science need to shake hands.

Richard Clarke Cabot (1868-1939)
The Meaning of Right and Wrong

Though penned decades ago, today these words ring more true than ever. Increasingly, physicians, researchers, policymakers, and the public face numerous quandaries brought on by advances in biology and medicine—advances the law is often ill-equipped to address. Years ago, a physician facing a frail newborn, such as "Baby Doe," had only to counsel the parents, perhaps call in a minister for additional family support, and offer the limited care then available until the child died. Now, such simplicity is nonexistent.

Scientific breakthroughs and novel medical technologies have led society to a point where pig and baboon livers have been transplanted into terminally ill persons (1-3), women with severe brain damage have been kept alive mechanically to continue their pregnancies (8), and research to help patients with Parkinson's disease has slowed due to a ban on federally funded protocols that use fetal tissue from induced abortions for transplantation (21). Even advances in medically assisted reproduction are simultaneously considered a blessing and immoral (20).

Often these dilemmas call for tragic choices, where two positive motivations are in conflict, such as a societal incentive to cut costs in the final weeks of life and the individual desire to keep a dying relative alive (19). Even when a legal imperative seemingly exists, more often than not decisionmaking blurs because of ethical and moral considerations. That is, while courts have issued decisions involving surrogate motherhood (e.g., the "Baby M" case), termination of life support (e.g., the cases of Karen Ann Quinlan and, more recently, Nancy Cruzan), and assisted suicide (e.g., the situation with Dr. Jack Kevorkian), ethical questions still remain.

Once generally the province of philosophy and religion, discussions about these and other highly complex and contentious issues have entered the political arena. During the past two decades, such discussions have crystallized into a discipline referred to, alternatively, as "biomedical ethics" or "bioethics." Today, how to set boundary rules for policy purposes amidst a web of ethical complexity has become extremely critical. **As the 21st century approaches, Congress faces policy dilemmas for which informed decisions require understanding bioethical considerations, as well as legal or economic dimensions. Furthermore, situations that demand ethical analysis are likely to arise with greater frequency and urgency.**

WHAT IS BIOETHICS?

The late 1960s and early 1970s marked the turning point in how Americans viewed medical innovation and biomedical research (16). Human subjects research in the United States reached its nadir with revelations of the Tuskegee syphilis study (box 1-A) and other abuses, such as the injection of liver cancer cells into patients at the Jewish Chronic Disease Hospital in Brooklyn, New York, and the intentional infection with hepatitis of residents of the Willowbrook State School for the Retarded (4,7,16). At about the same time, the term bioethics was coined by Van Rensselaer Potter, a cancer researcher in Madison, Wisconsin (15).

Originally, the word was envisioned as broadly encompassing an examination of the ethics of all biological sciences—e.g., ecology and agriculture (14)—not just biomedical research, medicine, and health care. With time, however, bioethics has become synonymous with biomedical ethics. **This report uses bioethics and biomedical ethics interchangeably, excluding areas of inquiry that others might include (e.g., environmental implications or the use of animals in experimentation).**

Bioethics evolved from the need to bring the perceived chaos of biology and medicine into the order of moral principle (12). Today, although most Americans might lack knowledge of bioethics as a discipline, the issues within its domain touch thousands each day; millions more are acquainted with them. Through mass media popularization, for example, few Americans are likely to be unfamiliar with the dilemmas raised by euthanasia or surrogate motherhood.

Bioethics is a field of study and a practice that involves professionals of many backgrounds—including philosophers, theologians, attorneys, clinicians, and researchers—who have a range of opinions; no one individual or profession can represent the breadth of perspectives that exist. Tens of thousands of individuals serve on Institutional Review Boards to review research involving human subjects or on institutional ethics committees in health care settings to consider ethical problems that arise in patient care (9). About 20 universities now offer degrees in bioethics (17), though the curricula vary widely (18). The growing importance of bioethics in society reflects both social change and the increased impact and complexity that advances in biology and medicine have brought to American life—advances that raise delicate policy questions.

ORIGINS AND ORGANIZATION OF THE REPORT

The social and cultural revolutions of the 1960s gave rise to the belief that government ought to seek resolution of issues raised by biomedical research and medical technology in a secular manner, consistent with pluralism. Ethical analysis has evolved as a useful tool for the evaluation and governance of new technologies, and biomedical ethics has been long of interest to

Box 1-A—The Tuskegee Syphilis Study

Just over 60 years ago, the U.S. Public Health Service (PHS) and several foundations began a study on approximately 600 African American males in Tuskegee, AL (5,6,11,13). At that time, the Tuskegee area had the highest incidence of syphilis in the Nation, and more than 400 of these men had this sexually transmitted disease, for which limited treatment was then available.

The men were lured into participating by the promise of free medical treatment, food, and burials. Initially, they were treated with mercury and arsenic compounds—then standard therapy—when the drugs were available. However, they also endured spinal taps without anesthesia and were denied penicillin long after it became apparent in 1945 that this antibiotic was the preferred drug. In fact, to prevent participants from receiving treatment by the U.S. Army, PHS also instructed draft boards not to induct them. Under congressional scrutiny, PHS officials offered the excuse that treating the subjects with penicillin would have arrested the disease and made following the long-term effects of syphilis impossible (11).

In 1972—four decades after it began—front page news reports brought this notorious abuse of human research subjects to an end (10). Elucidation of the Tuskegee syphilis study, along with other abuses in research involving human subjects (4,7,16), marked the start of bioethics' role in U.S. policy decisionmaking.

SOURCE: Office of Technology Assessment, 1993.

Congress (7,16). Federal interest in integrating bioethics into policy decisionmaking stems from a desire to understand the ethics surrounding Federal support for certain types of research, delivery of services in Federal programs, or payment for services in programs such as Medicare and Medicaid. The appeal of an institutionalized role, however, has waxed and waned over the past two decades. At present, no national policy forum exists for generally analyzing ethical issues associated with biological research and new medical technologies, though bioethicists testify before Congress and serve on Federal advisory committees.

In spring 1992, Congress signaled renewed interest in a formal role for bioethics in American governance during the Senate debate on reauthorization for the National Institutes of Health. Three factors stimulated this revived awareness and led to the congressional request for this background paper:

- concern about the many bioethical issues that have not been analyzed at the Federal level;
- the prospect that bioethical issues will arise with increased frequency and urgency in the future; and
- recognition that a national, institutional body to explore the role of biomedical ethics in U.S. public policy was nonexistent.

In September 1992, Senator Mark O. Hatfield, Ranking Minority, Committee on Appropriations; Senator Edward M. Kennedy, Chairman, Committee on Labor and Human Resources; and Senator Dennis DeConcini, Chairman, Subcommittee on Patents, Copyrights, and Trademarks, Committee on the Judiciary, asked the Office of Technology Assessment (OTA) to conduct a study that would assist Congress in determining possible approaches to examine policy problems with biomedical and ethical dimensions. That is, if Congress decides to create a new bioethics entity, what options should be considered?

Specifically, the congressional request sought a brief review of the history of broad-based Federal bioethics initiatives such as the National Commission for the Protection of Human Subjects of Biomedical and Behavioral Research

Table 1-1—Broad-Based Federal Policy Bodies in Biomedical Ethics[a]

Year	Initiative	Locus
1974-78[b]	National Commission for the Protection of Human Subjects of Biomedical and Behavioral Research	Department of Health, Education, and Welfare
1978-80[c]	Ethics Advisory Board	Department of Health, Education, and Welfare[d]
1978-83[e]	President's Commission for the Study of Ethical Problems in Medicine and Biomedical and Behavioral Research	Independent executive branch commission
1985-89[f]	Biomedical Ethics Advisory Committee	Congress

[a] Currently, as part of the Human Genome Project, the National Institutes of Health and the U.S. Department of Energy each fund grants through an Ethical, Legal, and Social Issues (ELSI) program. A joint ELSI Working Group advises both programs, but it is not a policy body.
[b] Public Law 93-348 (§ 202, 88 Stat. 342, 1974) created the National Commission in July 1974.
[c] Although disbanded in 1980, current DHHS regulations provide for the existence of an EAB (45 CFR 46.204). In fact, efforts to reestablish and recharter an EAB stalled in 1988 (53 FR 35232).
[d] With reorganization of the Department in 1980, EAB became part of the U.S. Department of Health and Human Services.
[e] Public Law 95-622 (42 U.S.C. Ch.6A) authorized creation of the President's Commission in November 1978 and set its termination for December 1982. Public Law 97-377 extended this date through March 1983. Due to delays in appointments and funding, the President's Commission was actually operational for just over 3 years.
[f] In reality, BEAC functioned for approximately 1 year. Public Law 99-158 established BEAC in May 1985. It was overseen by the Biomedical Ethics Board (BEB), which was comprised of Members of Congress. Almost a year elapsed before BEB was appointed and then nearly 2 1/2 more years before BEAC was constituted.

SOURCE: Office of Technology Assessment, 1993.

(National Commission), the President's Commission for the Study of Ethical Problems in Medicine and Biomedical and Behavioral Research (President's Commission), and the Biomedical Ethics Advisory Committee (BEAC), and asked OTA to analyze three questions based on this background material. First, what lessons could be learned from each of these entities? Second, what worked, or did not work? And finally, why?

To place these questions in context, chapter 2 briefly reviews the history of biomedical ethics in policy decisionmaking. In keeping with congressional interest and prerogative, the report focuses on the successes and failures of the three Federal bodies just mentioned, as well as the Ethics Advisory Board (EAB) of the Department of Health and Human Services. It also presents information on State and international initiatives; appendix A describes international bioethics bodies in greater detail. Appendix B presents the statutes or legislation establishing the National Commission, President's Commission, and BEAC, as well as the regulations that pertain to EAB and the charters under which it has operated. Federal bioethics initiatives devoted to a single issue are not analyzed in detail. Also excluded are the large number of academic bioethics centers, privately funded centers, and ethics committees of professional societies.

Chapter 3 examines the potential outlook for biomedical ethics in policy decisionmaking. It discusses what type of Federal effort might be created, factors to consider in launching a new body, and the role of Congress in such deliberations. The chapter is based, in part, on an OTA workshop convened in December 1992, in which participants from past Federal, State, and international bioethics forums, as well as individuals who observed (or used products of) these forums, discussed lessons from the past in light of future demands. As Senator Hatfield noted in opening the 1-day OTA workshop:

> . . . in public policy, if there is a vacuum, government eventually will fill it, right or wrong, good

> or bad. We just can't let difficult bioethical matters evolve at will, we ought to help direct them.

For, as then OTA Director John H. Gibbons pointed out, biology and medicine raise so many ethical issues of both a personal and public nature that:

> [in] a Nation that is extraordinarily more pluralistic, traditional authorities—for instance a single church or culture—can no longer provide guidance that will be acceptable to all. Thus, these issues come to rest with our government.

This report is intended to provide Congress with background material on what form a new Federal bioethics body could take and what might make a commission function effectively, not whether a commission should be established. Thus, discussions of specific bioethical dilemmas or an analysis of what issues are pressing are beyond the scope of this background paper.

CHAPTER 1 REFERENCES

1. Altman, L., "Terminally Ill Man Gets Baboon's Liver in Untried Operation," *New York Times*, June 29, 1992.
2. Altman, L., "Man Dying From Hepatitis Is Given a Baboon's Liver," *New York Times*, Jan. 11, 1993.
3. *Associated Press*, "Woman Gets Liver From a Pig But Dies," Oct. 13, 1992.
4. Beecher, H.K., "Ethics and Clinical Research," *New England Journal of Medicine* 274:1354-1360, 1966.
5. Caplan, A.L., "Twenty Years After: The Legacy of the Tuskegee Syphilis Study—When Evil Intrudes," *Hastings Center Report* 12(6):29-32, 1992.
6. Edgar, H., "Twenty Years After: The Legacy of the Tuskegee Syphilis Study—Outside the Community," *Hastings Center Report* 22(6):32-35, 1992.
7. Faden, R.R. and Beauchamp, T.L., *A History and Theory of Informed Consent* (New York, NY: Oxford University Press, 1986).
8. Fisher, M., "Germany's Fetal Position: If a Mother Dies, Should Her Pregnancy Continue?," *Washington Post*, Oct. 29, 1992.
9. Fletcher, J.C., "The Bioethics Movement and Hospital Ethics Committees," *Maryland Law Review* 50:859-894, 1991.
10. Heller, J., "Syphilis Victims in U.S. Study Went Untreated for 40 Years, *New York Times*, July 26, 1972.
11. Jones, J.H., *Bad Blood: The Tuskegee Syphilis Experiment* (New York, NY: Free Press, 1981).
12. Jonsen, A.R., "American Moralism and the Origin of Bioethics in the United States," *Journal of Medicine and Philosophy* 16:113-130, 1991.
13. King, P.A., "Twenty Years After: The Legacy of the Tuskegee Syphilis Study—The Dangers of Difference," *Hastings Center Report* 22(6):35-38, 1992.
14. Potter, V.R., *Bioethics: Bridge to the Future* (Englewood Cliffs, NJ: Prentice-Hall, 1971).
15. Reich, W.T., Georgetown University, Washington, DC, "How Bioethics Got Its Name," remarks at the conference, "The Birth of Bioethics," Seattle, WA, September 1992.
16. Rothman, D.J., *Strangers at the Bedside: A History of How Law and Bioethics Transformed Medical Decisionmaking* (New York, NY: Basic Books, Inc., 1991).
17. Schrage, M., " 'Bioethics' Burgeons, and Along With It Career Opportunities," *Washington Post*, Oct. 16, 1992.
18. Thornton, B.C., Callahan, D., and Nelson, J.L., "Bioethics Education: Expanding the Circle of Participants," *Hastings Center Report* 23(1):25-29, 1993.
19. U.S. Congress, Office of Technology Assessment, *Life-Sustaining Technologies and the Elderly*, OTA-BA-306 (Washington, DC: U.S. Government Printing Office, July 1987).
20. U.S. Congress, Office of Technology Assessment, *Infertility: Medical and Social Choices*, OTA-BA-358 (Washington, DC: U.S. Government Printing Office, May 1988).
21. U.S. Congress, Office of Technology Assessment, *Neural Grafting: Repairing the Brain and Spinal Cord*, OTA-BA-462 (Washington, DC: U.S. Government Printing Office, September 1990).

Lessons From the Past 2

In the absence of a single authoritarian church or other mechanisms to handle bioethical issues, American society often turns to government or the courts for resolution of thorny ethical issues. The reemerging interest in the role of bioethics in U.S. public policy signals the increasing importance of medical and biological technologies in daily life. The creation of Federal commissions stems from a desire for mechanisms to articulate common values and foster consensus in the face of growing cultural and religious heterogeneity. The need is not so much for finding moral solutions to complex policy matters, but rather, for identifying problems and either making recommendations or defining tradeoffs among alternatives.

FEDERAL INITIATIVES

Congress has exhibited an enduring interest in bioethics entities. Even before the term bioethics was coined in the early 1970s (73), the U.S. Senate deliberated in 1968 about a National Commission on Health Science and Society to examine "the social and moral" implications of biomedical advances (90). Since then, Congress has established three bodies to address ethical issues in medicine and research: the National Commission for the Protection of Human Subjects of Biomedical and Behavioral Research (National Commission), the President's Commission for the Study of Ethical Problems in Medicine and Biomedical and Behavioral Research (President's Commission), and the Biomedical Ethics Advisory Committee (BEAC). A fourth Federal initiative, the Ethics Advisory Board (EAB),[1] originated from a recommendation of the National Commission.

[1]Federal regulations (45 CFR 46) call for "Ethical Advisory Boards," and the original charter is for the "Ethical Advisory Board" (40). Soon after the board was constituted, however, it came to be referred to as the "Ethics Advisory Board"—a change also reflected in the second charter (9).

Box 2-A—Ethical, Legal, and Social Issues Programs, National Institutes of Health and U.S. Department of Energy

Since fiscal year 1988, Congress and the executive branch have made a commitment to determine the location on the DNA of all genes in the human body (e.g., as has been done for sickle cell anemia, cystic fibrosis, and Tay-Sachs disease) (84). The Human Genome Project is estimated to be a 15-year, $3-billion project. It has been undertaken with the expectation that enhanced knowledge about genetic disorders, increased understanding of gene-environment interactions, and improved genetic diagnoses can advance therapies for the 4,000 or so currently recognized human genetic conditions (15).

To address the ethical, legal, and social issues of the Human Genome Project, and to define options to address them, the National Institutes of Health (NIH) and the U.S. Department of Energy (DOE) each funds an Ethical, Legal, and Social Issues (ELSI) program. Funds for each agency's ELSI effort derive from a set aside of 3 to 5 percent of appropriations for the year's genome initiative budget. In fiscal year 1991, DOE-ELSI spending was $1.44 million (3 percent) and in fiscal year 1992 it was $1.77 million (3 percent); fiscal year 1993 spending is targeted at $1.87 million (22). NIH-ELSI spending for fiscal years 1990 and 1991 has been $1.56 million (2.6 percent) and $4.04 million (4.9 percent), respectively. For fiscal year 1992, NIH-ELSI spent $5.11 million (5 percent) and aims to spend $5.30 million in fiscal year 1993 (5 percent) (37).

ELSI funds bioethics research related to the Human Genome Project to expand the knowledge base in this area. The program operates in the model of peer review competition for grant funds. The ELSI Working Group, which advises both programs, initially framed the agenda and establishes priority research areas. Nevertheless, the nature of grant programs means the ultimate direction evolves from the bottom up—i.e., from the individual perspectives of researchers pursuing independent investigations—rather than from the top down—i.e., through policymakers or an overarching Federal body. Furthermore, no formal mechanism exists for ELSI-funded research findings to directly make their way back into the policy process (18,28,30,45,78). And although the ELSI programs have a large funding base for grants, they lack resources for in-house policy analysis. The ELSI Working Group, however, has played a role in policy analyses related to genetics and the Americans With Disabilities Act, cystic fibrosis carrier screening (88), and genetic research involving several family members (36).

Finally, although issues in human genetics are broad ranging, they comprise only a portion of bioethical issues. Because ELSI is the largest Federal funding source for bioethics studies, there is concern that a brain drain is occurring from nongenetic areas of bioethics to the ethics of human genetics research and applications (2).

SOURCE: Office of Technology Assessment, 1993.

Beyond these four bodies,[2] which had a general focus, bioethics has been a part of American governance via topic-specific initiatives, including: the Ethical, Legal, and Social Issues (ELSI) programs, National Institutes of Health (NIH) and U.S. Department of Energy (box 2-A), the NIH Human Fetal Tissue Transplantation Research Panel (Fetal Tissue Panel) (box 2-B), the Presidential Commission on the Human Immunodeficiency Virus Epidemic (Executive Order 12601; 52 FR 24129), the National Commission on Acquired Immune Deficiency Syndrome (Public

[2] In January 1993, NIH formally established a Science Policy Studies Center to advise the NIH Director on the ethical, legal, economic, and social implications raised by research; plans for staffing, establishing policies, and setting priorities are under development (1,43).

Box 2-B—The National Institutes of Health Human Fetal Tissue Transplantation Research Panel

Fetal tissue has long been used in research, including research involving transplanting fetal thymus tissue into humans (26,86). In 1975, the National Commission for the Protection of Human Subjects of Biomedical and Behavioral Research scrutinized ethical issues related to the use of fetuses in research (93) and developed the guidelines that were incorporated into Federal regulations for use of fetuses in research (45 CFR 46). Nevertheless, when scientists began using fetal tissue for neural grafting in the mid-1980s, questions were raised about the adequacy of these regulations for such research because they did not address specifically the therapeutic use of fetal tissue (86). The prospect that Federal funds would be used to support an intramural research protocol for implanting fetal neural tissue from induced abortions into persons with Parkinson's disease pushed matters to a head in 1987.

Lacking an Ethics Advisory Board within the Department of Health and Human Services to turn to, the Director of the National Institutes of Health (NIH) sought guidance from the Assistant Secretary for Health. In turn the Assistant Secretary directed NIH to appoint an ad hoc panel in March 1988, while simultaneously imposing a moratorium on Federal funding for the use of human fetal tissue from induced abortions for transplantation. The NIH Human Fetal Tissue Transplantation Research Panel, established as a subcommittee of the NIH Director's Advisory Committee, consisted of 21 members representing public interest, clinical, research, ethics, religious and legal perspectives (105).

The Panel's agenda was set by the Assistant Secretary through ten sets of ethical, social, legal, and technical questions. It met three times from September to December 1988 and, after considering material from invited speakers, public testimony, and commissioned papers (106), issued its report in December 1988 (105). The report concluded—on a 17 to 4 vote—that funding research involving the transplantation of human fetal tissue from induced abortions is acceptable public policy as long as carefully crafted safeguards are in place (105).

The recommendations were accepted unanimously by the NIH Director's Advisory Committee, which recommended the moratorium on fetal tissue transplantation research be lifted; the NIH Director concurred in a memorandum to the Assistant Secretary for Health in January 1989 (110). Despite these actions, none of the Panel's recommendations was implemented at that time. The Secretary extended the moratorium indefinitely in November 1989 (75), until it was lifted by President Clinton (58 FR 7468) when NIH was directed to develop guidelines based on the Panel's report in January 1993.

The ad hoc approach employed by the Federal Government clearly and publicly articulated the policy and ethical dimensions of fetal tissue transplantation and led to a specific recommendation, albeit with dissent; the process worked, although the recommendations were ignored by the initial client. However, the events leading up to the moratorium, and those that followed, raise questions of their own and add another layer of ethical considerations to the fetal tissue transplantation controversy: Is the Government's process for bioethical analysis adequate? And, what is the relationship between personal ethical convictions and the appropriate shape of public policy in a pluralistic society?

SOURCE: Office of Technology Assessment, 1993.

Law 100-607; § 241-249, 102 Stat. 4223, 1988), and the U.S. Department of Health and Human Services' (DHHS) Organ Transplantation Task Force (Public Law 98-507). OTA reports also have considered the bioethical dimensions of a range of issues (80-89). Additionally, bioethical considerations have been included as part of the deliberations about gene therapy by the NIH Recombinant DNA Advisory Committee (RAC), though the focus and function of RAC and the

AIDS commissions are less bioethics per se than either the ELSI programs or the Fetal Tissue Panel. The Ethics and Values in Science and Technology Program of the National Science Foundation once supported analyses of ethical issues related to specific issues arising from research, but avoided aspects related to clinical care because they were beyond the agency's mission. Although not involved in policy development, the National Endowment for the Humanities has also supported projects in bioethics—typically courses, book projects, or workshops.

This section briefly reviews the history of the four principal, broad-based Federal initiatives: the National Commission, EAB, President's Commission, and BEAC; it discusses their creation, procedures, and products. Chapter 3 analyzes these practices and processes in the context of defining common elements to elucidate alternative Federal forums to integrate biomedical ethics in U.S. public policy. For additional detail, appendix B contains the statutes or legislation establishing the National Commission, President's Commission, and BEAC, as well as the regulations and charters that pertain to the EAB.

■ National Commission for the Protection of Human Subjects of Biomedical and Behavioral Research

The National Research Act (Public Law 93-348 (§ 202, 88 Stat. 342, 1974) created the National Commission in July 1974, after earlier attempts at constituting a similar commission failed (23,74,90). In establishing the National Commission, Congress directed it to identify the principles of ethics needed to protect human subjects involved in research and to use those principles to recommend actions by the Federal Government.

Eleven members were appointed by the Secretary of the then Department of Health, Education, and Welfare (DHEW): five scientists, three lawyers, two ethicists, and one person in public affairs (111); appointments were for the full term of the commission—4 years. Of the five scientists, three were physicians and two were psychologists (111). During its existence, commissioners met on nearly a monthly basis (111). Within the first year, 16 staff were hired, and in May 1975, the National Commission issued its first report, which addressed research involving fetuses (93). By July 1975, this report's recommendations had been translated into Federal regulations. The startling speed with which change was effected was brought about by a clause in the law that forced the Secretary to accept the National Commission's recommendations or make public the reasons for rejection. The clause, however, seemed to have an effect only as long as the National Commission was operative. After it disbanded, DHEW began to ignore the forcing clause for recommendations of later reports (38,39).

Ten reports and many appendixes followed the initial effort (94-102); some as successful, others not. For example, the National Commission's reports on ethical guidelines to protect certain classes of human subjects in research—fetuses, prisoners, and children (93,94,97)—led to Federal regulations (45 CFR 46), and today an NIH office oversees their enforcement (61). The National Commission identified the basic ethical principles to be applied in the ethical evaluation of human subjects research (98), which was also codified (45 CFR 46). More controversial and largely ignored (16,111) was the National Commission's work on psychosurgery (96). Its work on research and the "institutionalized mentally infirm" (101) was never implemented; regulations were proposed after the National Commission's demise, but never finalized—in violation of the law (38,39).

The National Commission brought its work to a close in 1978, but spawned the next broad-based Federal entity, the EAB—again through the forcing clause. Rather than publish objections within 180 days as to why a continuing Federal body should not be established, DHEW incorporated the National Commission's recommendations into its regulatory framework (45 CFR 46.204) and established the EAB.

Ethics Advisory Board

As early as 1970, NIH discussed the need for a body to advise the Secretary and DHEW on controversial ethical, legal, and social issues posed by biomedical research protocols (23,24). Nevertheless, the EAB was not established until 1978, following a recommendation of the National Commission.

The Secretary appointed an 11-member Board that included lawyers, a theologian, a philosopher, clinicians, researchers, and a member of the public. Initially, the Board had eight staff, as well as consultants and student assistants. During its approximately 2-year existence, it met approximately 20 times (42). In vitro fertilization (IVF) was the first topic addressed by EAB, and its 1979 report stipulated several criteria to be met for approval of federally funded research in this area. Among the report topics that followed were a report on fetoscopy and items related to Freedom of Information Act inquiries (41). In all, EAB produced four documents (91,92,103,104).

Although Federal regulations define EAB's scope to issues involving the fetus, pregnant women, and human IVF (45 CFR 46.201), the original charter under which the Board operated clearly defines EAB's scope much more broadly as a standing body to review ethical issues of biomedical research (40); the scope and level of activities were further widened with the subsequent charter (9). And in fact, the Secretary used EAB in a broad manner to report on ethical issues raised by research unrelated to the three specified activities (41,103,104).

In contrast to the other three Federal initiatives, EAB was intended as an ongoing, standing board with a mission to examine issues related to specific protocols or types of research as they arose—a logical notion given the quickening pace of biomedical research. Additionally, Federal regulation required an EAB review prior to funding research on human IVF (45 CFR 46.204d). Nevertheless, despite the regulatory requirement for an EAB (45 CFR 46.204), DHHS disbanded it in 1980 at the direction of the Office of Science and Technology Policy (35), and thus violated its own regulations (77). The appearance of the President's Commission in 1978 contributed to EAB's demise because policymakers failed to distinguish their distinct purposes. Through its broad charter, EAB was positioned to examine research protocols that raised novel issues and to devise procedures and criteria for their review and implementation. In contrast, the President's Commission was a forum for national debate on global issues of bioethical concern.

In 1988, OTA's report on medical and social issues of infertility (84) forced the debate over DHHS's failure to maintain an EAB to the surface (77,79). Federal funding for peer-reviewed, approved projects was clearly blocked without an EAB.[3] DHHS promised to reestablish the Board and published a proposed charter for a new EAB in 1988 (53 FR 35232). The new charter called for an expanded membership of 21 individuals—drawn from specific, but diverse fields of expertise—to serve for overlapping 4-year terms. Meetings were to take place approximately 10 times annually. The comment period generated nearly 200 signatories to various positions—with a clear majority in favor, although with caveats about the frequency of meetings, number of individuals, and other details. A revised charter was drafted, but never signed in the waning days of the Reagan Administration, and no EAB materialized during the Bush administration.

[3] NIH estimated this de facto ban on federally funded research related to human IVF was such that more than 100 grant applications in this area were not submitted between 1980 and 1987 because of a widespread awareness that while such grants might be approved, they would go unfunded because no EAB would exist to review them (27,84).

President's Commission for the Study of Ethical Problems in Medicine and Biomedical and Behavioral Research

Congress created the President's Commission in 1978 (Public Law 95-622; 42 U.S.C. Ch.6A). As just mentioned, amid confusion that the President's Commission's mandate overlapped the EAB's, Congress convinced DHEW to divert its appropriations to support the new body (18). And though funding for the President's Commission eventually came from other sources (13), EAB remains dormant. Congress also elevated the new body to independent presidential status, in contrast to the National Commission, which had operated autonomously within DHEW. The scope was extended beyond human subjects research to include medical practice, and the commission was granted broad authority to expand beyond the seven mandated topics to report on emerging issues on its own initiation or at the request of the President or the head of an agency.

Appointment powers resided with the President. By summer 1979, the 11 commissioners had been appointed for rotating terms, and the first meeting was held in January 1980. By law, commissioners were drawn from specific areas: three who practiced medicine, three biomedical or behavioral researchers, and five from other fields. Over the President's Commission's duration, this latter category included individuals from law, sociology, economics, and philosophy, as well as a homemaker and a businessman. In all, 21 different commissioners served on the President's Commission. The body was well staffed: During the 3 years the President's Commission functioned, about 30 to 40 people worked for it, but generally only 20 at any given time.

Like the National Commission, the legislation creating the body also had a forcing clause. But unlike the National Commission, the reports of the President's Commission—as a matter of explicit policy—made few specific recommendations (108) to which agencies needed to respond. Instead, the President's Commission produced consensus reports that largely articulated mainstream views (108) on the mandated topics, as well as three additional reports not requested in the original legislation; a summary document of the Commission's work was also published, as well as several appendixes and proceedings (62-72). These documents were highly regarded and many have had sustained policy influence (109).

For example, its report on the definition of death became the foundation for statutory changes adopted throughout the Nation (12). Its report on foregoing life-sustaining treatment—undertaken on the Commission's own initiative—was probably the most influential (7,13,35) and remains an important point of reference for courts and legislatures. The Commission's report on recombinant DNA research (66) led the NIH RAC to establish a working group to consider both technical and ethical aspects of human gene therapy. On the other hand the report on health care access was less influential (4,6,20,21). Still others suffered inattention at the time of their release—e.g., the report on genetic screening and genetic counseling and the report on whistleblowing in biomedical research (13)—but were remarkably prescient about issues that surfaced in the 1990s (8,88,112).

After one 3-month extension for its authority, the President's Commission expired in March 1983. Its recommendation that a similar body be created on its termination became the focus of almost immediate attention (18), thus setting the stage for the most recent congressional sortie into institutionalizing biomedical ethics.

Biomedical Ethics Advisory Committee

In May 1985, Congress looked to itself to house the fourth, and most recent, Government-sponsored bioethics body: BEAC (Public Law 98-158). With the President's Commission's sunset in March 1983, Congress repeatedly expressed interest in reconstituting some type of bioethics commission (16-18,76). Overseeing BEAC was the Biomedical Ethics Board (BEB), modeled on the Technology Assessment Board that oversees OTA: 12 Members of Congress equally divided by chamber and political affiliation.

Nearly a year passed before the House and Senate leaderships appointed Board members, who in turn were charged with appointing a 14-member BEAC. Two lay members and representatives from law, ethics, biomedical research and clinical care were appointed—2 1/2 years later. Less than 1 week before it was scheduled to expire, BEAC held its first meeting in September 1988.

Two full-time staff worked for BEAC, which initially was to analyze three mandated topics: human genetic engineering (i.e., gene therapy), fetal research, and feeding and nutrition of dying patients (18). To address the first topic, it held its second meeting in February 1989. Shortly thereafter, however, Senate BEB members deadlocked on choosing a chairman along partisan, prochoice-antiabortion lines (18). BEAC's proposed budget—sufficient for 12 staff to address the mandated topics—was cut and spending made contingent on a fully constituted BEAC. BEAC expired in September 1989 having issued no reports.

STATE ENTITIES

Although bioethics forums were initially confined to federally funded efforts, more recently, State legislatures and executive branches have begun formal efforts to incorporate bioethics in their analytic and decisionmaking processes (34). Most State panels have been devoted to a single issue, particularly health care access (e.g., California (14), Minnesota (11), Oregon (25), Vermont (107)). The Minnesota House of Representatives created a bioethics subcommittee and held hearings during its 1991-92 session (10). At least three States—New Jersey, New York, and Colorado—created entities designed to consider a broad range of issues, though New Jersey's commission currently is unfunded and Colorado's effort has not yet been funded (13,44).

■ New Jersey Commission on Legal and Ethical Problems in the Delivery of Health Care

New Jersey's efforts to address bioethical issues developed from a series of landmark decisions by the State Supreme Court, beginning in 1976 with the "Karen Ann Quinlan case"—the first case to address refusal of life-sustaining treatment (32). In 1985, the court dealt with a proposal to withdraw a feeding tube from a debilitated and demented elderly nursing home patient (31). In both cases, the court stated that the opinions were not intended to set guidelines for life-sustaining treatment decisions and that these issues are more suitably addressed by the legislative process, which can accommodate the different needs and interests represented in New Jersey's communities.

In November 1985, the State legislature created the New Jersey Commission on Legal and Ethical Problems in the Delivery of Health Care (New Jersey Bioethics Commission) as a permanent legislative commission (New Jersey Public Law 1985, Chapter 363). Though currently unfunded, it operated for 5 years with the mandate to "provide a comprehensive and scholarly examination of the impact of advancing technology on health care decisions [in order to] enable government, professionals in the fields of medicine, allied health care, law, and science, and the citizens of New Jersey and other States to better understand the issues presented, their responsibilities, and the options available to them."

The New Jersey Bioethics Commission was comprised of a diverse, multidisciplinary group of 27 appointed members. Commissioners were drawn from a broad spectrum of expertise and opinions, including medicine, nursing, health care administration, natural science, social science, law, the humanities, theology, ethics, and public affairs. By law, the Commission included representatives of the State legislative and executive branches, major professional and health care associations, and professional and public communities. Fourteen at-large members were appointed by the Governor, the Senate President, and the Speaker of the General Assembly. One Commissioner represented the Citizens' Committee on Biomedical Ethics—demonstrating the trend toward grassroots organizations in bioethics and health care decisionmaking (19,29,34).

The Commission's statutory mandate was broad, and it enjoyed substantial freedom to set its own agenda. Areas addressed included: determination of death, advance directives, and decisionmaking for incompetent patients without advanced directives (46-48,50,51). Following the decision in the "Baby M case" (33), the Commission undertook a study of surrogate motherhood (49).

During its approximately 6-year existence, fiscal support varied widely. Staffing ranged from five full-time professionals and two part-time consultants to two full-time professionals and two consultants (3,60). Staff to the Commision had broad freedom to consult outside experts, select papers for presentation to the Commission, hold public hearings, and do empirical research on life-sustaining treatment, determination of death, and reproductive issues. At least one former staff member, however, believes the Commissioners and staff did not have adequate access to the Governor or ongoing cooperation from the Governor or the Legislature (5). Conflicts between staff and commissioners over substantive and nonsubstantive issues also hampered some deliberations (5).

The New Jersey Bioethics Commission conducted its deliberations publicly, believing that openness to public participation and scrutiny was necessary if the Commission was to be responsive and credible; it also held numerous public hearings. At times, the meetings were highly politicized. In particular, the four legislators on the Commission were often divided; three didn't approve of the Commission's existence and spent much of their time trying to disband the Commission rather than participate in discussions on a particular issue (5). On the other hand, including elected representatives and executive branch officials established an important liaison to those who ultimately effect change; it worked well sometimes, but was obstructionist on other occasions (3,60).

The Commission also created five ad hoc task forces for detailed studies of new reproductive practices, institutional ethics committees, public and professional education, AIDS, and protection of vulnerable patients. Each task force consisted of 12 to 20 members, including both Commissioners and others selected for their specialized expertise. Task Forces made recommendations to the Commission, which retained authority to accept, reject, or modify the proposals.

For each of the topics it studied, the New Jersey Bioethics Commission, sometimes jointly with a task force, published reports designed to explain "the intent and spirit of the recommendations" and to "enhance understanding and promote discussion of bioethical issues by policymakers, members of the legal and health care communities and by all New Jersey citizens" (48). The Commission published six documents, ranging from comprehensive policy analyses to a guidebook for health professionals to consumer-targeted information documents (46-51), one of which also was published in Spanish (47). Commission work in several areas resulted in specific State laws, including: the New Jersey Declaration of Death Act (New Jersey Public Law 1991, Chapter 90) and the New Jersey Advance Directives for Health Care Act (New Jersey Public Law 1991, Chapter 201).

■ New York State Task Force on Life and the Law

In the early 1980s, New York State faced a mounting crisis over "Do Not Resuscitate" (DNR) orders, directives that advise physicians whether to resuscitate a patient. A grand jury investigation revealed widespread abuses associated with DNR orders in health care facilities, and the grand jury's findings and accompanying public outcry helped lead to the establishment in 1984 of the New York State Task Force on Life and the Law. The Task Force still functions today, receiving funds on an annual basis. Designed to provide counsel on a broad range of topics—e.g., surrogate parenting, determination of death, and physician assisted suicide—it makes policy recommendations to the State executive and legislative branches; its agenda is established in consultation with the Governor and New York State Department of Health (44).

The Governor appoints all Task Force members, who include doctors, nurses, and representatives of different religious communities and public interest groups. Members are chosen from both political parties and to reflect different perspectives within the State. The role of the Task Force is purely advisory; it is not involved with final policy determination. The group was set up as an independent entity, not as a division of an existing department. Nevertheless, by factoring in the views of representatives from various State communities, it has been able to identify points of consensus and recommend proposals that were acceptable to the legislature, State agencies, and the Governor (44).

In addressing a topic, the Task Force examines existing literature and takes into account the range of political and social concerns in the State to create generalized guidelines that fit the vagaries of New York State's legal and cultural climate (44). The Task Force has been well staffed and is further aided by consultants, who participate on a pro bono basis.

Because the Task Force is advisory, it has been exempt from open meeting laws that apply to other State bodies. To date, its meetings have not been open to the public. Nor has the Task Force held public hearings, although after recommendations have been made, hearings have been held as part of the legislative process. One former staff member believes that the ability to conduct closed meetings has contributed to the Task Force's success—i.e., that private deliberations insulate members from political pressures that can surround issues under consideration (44).

The Task Force has produced eight reports supporting its recommendations (52-59). Most included legislative or regulatory proposals, and all recommendations were drafted and enacted in some form. Topics addressed include: organ and tissue transplantation, determination of death, health care agents, surrogate parenting, and surrogate decisionmaking for incapacitated patients. Reflecting the difficulty of achieving consensus, abortion has not been a Task Force issue.

INTERNATIONAL EFFORTS

In the past few years, bioethics has become a global enterprise (table 2-1). Hospitals the world over have established ethics committees, and many academic and professional bioethics forms have been created in other countries. The governments of at least 27 nations on 6 continents have established national commissions of some type or currently have legislation pending. Thus, while U.S. Government-sponsored bioethics forums have disappeared, government initiatives are on the rise elsewhere. Multinational organizations have also begun to analyze bioethical issues through committees or commissions. Appendix A chronicles these and individual country efforts; this section summarizes some common themes and highlights some differences.

Not surprisingly, the purposes of bodies in other countries vary widely. Some advise parliaments directly, others exist to stimulate and educate the public. Still others assume the role of distilling and articulating the country's sensibility on bioethics matters.

Each country integrates biomedical ethics into its policymaking structure in a different manner, and no single approach predominates. Unique cultural aspects are key and influence the development of bioethical approaches in a particular country; what might be viewed as acceptable in one country could be unethical in another. Until

Table 2-1—Typology of International Bioethics Commissions[a,b,c]

	National commission	Other government commission	Hospital	Professional	Academic	Private
Argentina	1[d]		Y		Y	Y
Australia	1	e	Y	Y	Y	Y
Austria[f]			Y		Y	
Belgium	d	e,g	Y	Y	Y	
Botswana						Y
Brazil	1		Y	Y	Y	
Canada	2	e,g	Y	Y	Y	
Chile			Y	Y	Y	Y
China	d	g			Y	Y
Columbia[h]		i		Y	Y	Y
Croatia		i	Y		Y	
Cyprus[f]	d			Y		
Czech Republic	1	e	Y			
Denmark	2	e	Y		Y	
Egypt					Y	
Finland	2				Y	
France	1		Y	Y	Y	Y
Germany[j]		g,l		Y	Y	
Ghana[k]				Y	Y	
Greece	1		Y	Y	Y	Y
Holy See[f]		l	Y			
Hungary	2		Y	Y	Y	
Iceland			Y	Y		
Ireland[f]			Y			
Israel	2			Y	Y	Y
Italy	2	e,g	Y	Y	Y	Y
Japan		g	Y		Y	Y
Kuwait[m]					Y	
Liechtenstein[f,n]						
Luxembourg	1					
Malta	1	g			Y	
Mexico	1		Y		Y	Y
The Netherlands	2		Y	Y	Y	Y
New Zealand	1[d]	g,o	Y	Y	Y	
Norway	3				Y	Y
Peru			Y	Y		
The Philippines[p]	1		Y		Y	
Poland	3		Y		Y	
Portugal	1		Y	Y	Y	
Romania			Y	Y	Y	
Russia	1				Y	Y

recently, most bioethics commissions in other countries have been temporary bodies devoted to one or a small number of topics selected in advance by the sponsor. In 1983, however, France created a broad-based bioethics commission, and since then several other European nations have followed suit.

In contrast to the United States, many of the national commissions abroad limit public access, and meetings are generally closed; in some cases, members of the public may offer their views through periodic public symposia. As in the United States, however, most governments strive to include membership of non-health care professionals; in some cases, physicians and scientists are a clear minority—such as on the 17-member Danish Council of Ethics, where laypersons represent the majority.

	National commission	Other government commission	Hospital	Professional	Academic	Private
Scotland[q]			Y			
South Africa		r		Y	Y	
Spain		g,l	Y	Y	Y	Y
Sweden	1	g	Y	Y	Y	Y
Switzerland		g	Y	Y	Y	Y
Turkey	1[d]		Y	Y	Y	
United Kingdom		g	Y	Y	Y	Y
United States		e,g	Y	Y	Y	Y
Uruguay			Y	Y	Y	

[a] Unless otherwise indicated, information is based an OTA survey of international government officials and bioethics experts. However, because survey responses varied widely, this table likely represents an incomplete picture of activities in the area of hospital, professional academic, and private entities. Hence, conclusions should not be drawn about absences for any particular entry.
[b] Numbers under the "National commission" column indicate how many federally sponsored entities currently exist to examine either bioethical issues, generally, or research-related issues, specifically. "Hospital" refers to hospital ethics committees or research ethics boards. "Professional" refers to subcommittees within or between professional groups, such as medical or nursing associations. "Academic" refers to departments or curricula in medical schools or other academic institutions. "Private" refers to private organizations devoted to the study of bioethics, excluding professional or academic groups. "Y" indicates these activities in a country.
[c] Though no all-encompassing international organization exists, several multinational organizations (e.g., the Council of Europe and the Commission of the European Community) have issued policy statements and sponsored forums for discussing bioethical issues (app. A). In addition, regional bioethics commissions and/or regional private groups have begun to collectively organize (e.g., in Central/Eastern Europe, Scandinavia, and Latin America).
[d] Legislation is pending to create a new, or additional, national bioethics commission.
[e] Local or regional commissions have existed.
[f] S. Le Bris, "National Ethics Bodies," contract document for the Council of Europe, Ad Hoc Committee of Experts on Bioethics, Round Table of Ethics Committees, Madrid, Spain, Mar. 24, 1992.
[g] Ad hoc, topic-specific commissions have existed.
[h] F. Sanchez-Torres, "Background and Current Status of Bioethics in Columbia," *Bioethics: Issues and Perspectives*, S.S. Connor and H.L. Fuenzalida-Puelma (eds.) (Washington, DC: Pan American Health Organization, 1990).
[i] Laws addressing specific bioethical issues have been enacted.
[j] H.-M. Sass, "Blue-Ribbon Commissions and Political Ethics in the Federal Republic of Germany," *Journal of Medicine and Philosophy* 14:465-472, 1989; and D. Wikler and J. Barondess, "Bioethics and Anti-Bioethics in Light of Nazi Medicine: What Must We Remember," *Kennedy Institute of Ethics Journal* 3:39-55, 1993.
[k] J. Hevi, "In Ghana, Conflict and Complementarity," *Hastings Center Report* 19(4):S5-S7, 1989.
[l] Religious leaders have formed committees to discuss bioethical issues.
[m] M. Al-Mutawa, "Health Care Ethics in Kuwait," *Hastings Center Report* 19(4):S11-S12, 1989.
[n] No ethics bodies exist in this country.
[o] The government has commissioned reports from individuals.
[p] L.D. de Castro, "The Philippines: A Public Awakening," *Hastings Center Report* 20(2):27-28, 1989.
[q] D.M. Tappin and F. Cockburn, "Ethics and Ethics Committees: HIV Serosurveillance in Scotland," *Journal of Medical Ethics* 18:43-46, 1992.
[r] The government has issued guidelines and reports in bioethics, but not through commissions.

SOURCE: Office of Technology Assessment, 1993.

WHAT WORKED?

Each past effort existed in unique circumstances that contributed to its success or failure. It's largely acknowledged that three of the four Federal efforts succeeded. However, the most recent endeavor, BEAC, failed. New Jersey and New York approached biomedical ethics with different approaches, but each State's effort worked for its own jurisdiction. Commissions abroad offer a rich array of options to evaluate, although until recently most have been topical. France, Denmark, and the United Kingdom, however, have well-developed, wide-ranging efforts, and all have their strengths.

What generally made bioethics bodies successful? Many factors—tangible and intangible—contributed to success. Timing and personalities were important, but were difficult to predict beforehand. A few themes, nevertheless, persisted across the success stories and were notably absent in the failure. **As elaborated in the next chapter, adequate staffing and funding improved the chances for success. Successful commissions were relatively free of political interference, had flexibility in addressing issues, were open in their process and dissemination of findings, and were comprised of a diverse group of individuals who were generally free of ideology and had wide ranging expertise.**

Individually examining past Federal bioethics commissions reinforces this assessment. The National Commission was formed in response to a critical need to resolve several biomedical research issues that had accumulated over time. Its appearance was well-timed, and it was well staffed. Its placement within the agency it would guide—the U.S. Department of Health, Education, and Welfare—and requiring this agency to respond to its recommendations were critical to the National Commission's subsequent contribution in ensuring ethical conduct in federally funded research involving humans. The EAB's structure allowed for a flexible agenda to respond to the biomedical research community's pressing needs for guidance, and as a standing unit it also could respond quickly. The President's Commission was able to distance itself from political influence, was adequately funded and well staffed, and received a broad mandate. In contrast, BEAC and its congressional board suffered from insufficient staffing, political conflict, and excessive debate over its agenda.

Lessons from State efforts can also be drawn, though New York provides a less useful model because it holds closed deliberations—a process that would be illegal for any new Federal effort presumably subject to the sunshine provisions of the Federal Advisory Committee Act that require open meetings (5 U.S.C. Ap. 2, § 1 et seq.). Still, both bodies in New Jersey and New York were well staffed, had leeway to consult experts outside the commission, and had flexible agendas.[4] In contrast, international commissions are poorly staffed compared to U.S. efforts and most, like New York, hold closed meetings. Public funding for outreach in Denmark, the wide media and government attention paid to France's effort, and the unusual cooperation of the U.K. Government with its private council are striking and undoubtedly contribute to their success.

CHAPTER 2 REFERENCES

1. American Association for the Advancement of Science, "NIH Establishes New Center for Science Policy Studies, *Professional Ethics Report* 6(1):1.
2. Annas, G.J. and Elias, S., "Social Policy Research Priorities for the Human Genome Project," *Gene Mapping: Using Law and Ethics as Guides*, G.J. Annas and S. Elias (eds.) (New York, NY: Oxford University Press, 1992).
3. Armstrong P.W., Bridgewater, NJ, personal communication, March 1993.
4. Arras, J.D., "Retreat From the Right to Health Care: The President's Commission and Access to Health Care," *Cardozo Law Review* 6:321-346, 1984.
5. Asch, A., Boston University of School of Social Work, Boston, MA, remarks at "Biomedical Ethics in U.S. Public Policy," a workshop sponsored by the Office of Technology Assessment, U.S. Congress, Dec. 4, 1992.
6. Bayer, R., "Ethics, Politics, and Access to Health Care: A Critical Analysis of the President's Commission for the Study of Ethical Problems in Medicine and Biomedical and Behavioral Research," *Cardozo Law Review* 6:303-320, 1984.
7. Brock, D.W., Brown University, Providence, RI, personal communication, March 1993.
8. Burd, S., "Public Health Service Plans Hearing for Scientists Accused of Fraud," *Chronicle of Higher Education*, vol. 38, June 17, 1992.
9. Califano, J.A., Secretary of Health, Education, and Welfare, "Charter: Ethics Advisory Board," Jan. 11, 1979.

[4] New York still operates under these conditions. The Task force does, however, hold public hearings once recommendations have been put forth.

10. Caplan, A.L., University of Minnesota, Minneapolis, MN, personal communication, March 1993.
11. Caplan, A.L. and Ogren, P.A., "Do the Right Thing: Minnesota's HealthRight Program," *Hastings Center Report* 22(5):4-5, 1992.
12. Capron, A.M., "Looking Back at the President's Commission, *Hastings Center Report* 13(5):7-10, 1983.
13. Capron, A.M., University of Southern California Law Center, Los Angeles, CA, remarks at "Biomedical Ethics in U.S. Public Policy," a workshop sponsored by the Office of Technology Assessment, U.S. Congress, Dec. 4, 1992.
14. Colbert, T., "Public Input Into Health Care Policy: Controversy and Contribution in California," *Hastings Center Report* 20(5):21, 1990.
15. Cook-Deegan, R.M., "The Human Genome Project: The Formation of Federal Policies in the United States, 1986-1990," *Biomedical Politics*, K.E. Hanna (ed.) (Washington, DC: National Academy Press, 1991).
16. Cook-Deegan, R.M., "Finding a Voice for Bioethics in Public Policy: Federal Initiatives in the United States, 1974-1991," unpublished manuscript, November 1992.
17. Cook-Deegan, R.M., Institute of Medicine, Washington, DC, remarks at "Biomedical Ethics in U.S. Public Policy," a workshop sponsored by the Office of Technology Assessment, U.S. Congress, Dec. 4, 1992.
18. Cook-Deegan, R.M., *The Gene Wars: Science, Politics, and the Human Genome* (New York, NY: W.W. Norton, forthcoming 1993).
19. Coombs, B.J., "Community Values and Health Care Costs," *Washington Post*, Feb. 15, 1993.
20. Daniels, N., *Just Health Care* (New York, NY: Cambridge University Press, 1985).
21. Daniels, N., *Am I My Parents Keeper? An Essay on Justice Between the Young and the Old* (New York, NY: Oxford University Press, 1988).
22. Drell, D., Human Genome Program, Office of Health and Environmental Research, U.S. Department of Energy, Germantown, MD, personal communication, February 1993.
23. Faden, R.R. and Beauchamp, T.L., *A History and Theory of Informed Consent* (New York, NY: Oxford University Press, 1986).
24. Fletcher, J.C., University of Virginia, Charlottesville, VA, personal communication, January 1993.
25. Garland, M.J. and Hasnain, R., "Health Care in Common: Setting Priorities in Oregon," *Hastings Center Report* 20(5):16-18, 1990.
26. Hansen, J.T. and Sladek, J.R., Jr., "Fetal Research," *Science* 246:775-779, 1989.
27. Haseltine, F.P., National Institute of Child Health and Human Development, Bethesda, MD, personal communication, Sept. 25, 1987.
28. Healy, B.P., National Institutes of Health, testimony, *Domestic and International Data Protection Issues*, hearings before the Subcommittee on Government Information, Justice, and Agriculture, Committee on Government Operations, U.S. House of Representatives, Oct. 17, 1991 (Washington, DC: U.S. Government Printing Office, 1991).
29. Hill, T.P., "Giving Voice to the Pragmatic Majority in New Jersey," *Hastings Center Report* 20(5):20, 1990.
30. Hubbard, R. and Wald, E., *Exploding the Gene Myth* (Boston, MA: Beacon Davis, 1993).
31. *In re Conroy*, 98 N.J. 321, 486 A.2d 1209 (1985).
32. *In re Quinlan*, 70 N.J. 10, 355 A.2d 647, *cert. denied sub nom. Garger* v. *New Jersey*, 429 U.S. 922 (1976).
33. *In the Matter of Baby M*, 109 N.J. 396, 537 A.2d 1127 (1988), *affirming in part, reversing in part*, 217 N.J. Super. 313, 525 A.2d 1128 (1987).
34. Jennings, B., "Grassroots Bioethics Revisited: Health Care Priorities and Community Values," *Hastings Center Report* 20(5):16, 1990.
35. Jonsen, A.R., University of Washington, Seattle, WA, personal communication, March 1993.
36. Juengst, E.T., National Center for Human Genome Research, National Institutes of Health, Bethesda, MD, personal communication, March 1993.
37. Langfelder, E.J., National Center for Human Genome Research, National Institutes of Health, Bethesda, MD, personal communication, February 1993.
38. Levine, R.J., *Ethics and Regulation of Clinical Research* (2nd ed.) (Baltimore, MD: Urban and Schwarzenberg, 1986).

39. Levine, R.J., Yale School of Medicine, New Haven, CT, personal communication, March 1993.
40. Matthews, D., Secretary of Health, Education, and Welfare, "Charter: Ethical Advisory Board," Dec. 27, 1976.
41. McCarthy, C.R., Kennedy Institute of Ethics, Washington, DC, remarks at "Biomedical Ethics in U.S. Public Policy," a workshop sponsored by the Office of Technology Assessment, U.S. Congress, Dec. 4, 1992.
42. McCarthy, C.R., Kennedy Institute of Ethics, Washington, DC, personal communication, March 1993.
43. Mervis, J., "NIH Forms Policy Centre to Study Research Ethics," *Nature* 352:367, 1992.
44. Moskowitz, E.H., The Hastings Center, Briarcliff Manor, NY, remarks at "Biomedical Ethics in U.S. Public Policy," a workshop sponsored by the Office of Technology Assessment, U.S. Congress, Dec. 4, 1992.
45. Murray, T.H., "Speaking Unsmooth Things About the Human Genome Project," *Gene Mapping: Using Law and Ethics as Guides*, G.J. Annas and S. Elias (eds.) (New York, NY: Oxford University Press, 1992).
46. New Jersey Commission on Legal and Ethical Problems in the Delivery of Health Care, *Problems and Approaches in Health Care Decision-making: The New Jersey Experience* (Trenton, NJ: State of New Jersey Commission on Legal and Ethical Problems in the Delivery of Health Care, 1990).
47. New Jersey Commission on Legal and Ethical Problems in the Delivery of Health Care, *Advance Directives for Health Care: Planning Ahead for Important Health Care Decisions* (Trenton, NJ: State of New Jersey Commission on Legal and Ethical Problems in the Delivery of Health Care, 1991).
48. New Jersey Commission on Legal and Ethical Problems in the Delivery of Health Care, State of New Jersey, *The New Jersey Advance Directives for Health Care and Declaration of Death Acts: Statutes, Commentaries and Analyses* (Trenton, NJ: State of New Jersey Commission on Legal and Ethical Problems in the Delivery of Health Care, 1991).
49. New Jersey Commission on Legal and Ethical Problems in the Delivery of Health Care, State of New Jersey, *After Baby M: The Legal, Ethical and Social Dimensions of Surrogacy* (Trenton, NJ: State of New Jersey Commission on Legal and Ethical Problems in the Delivery of Health Care, 1992).
50. New Jersey Commission on Legal and Ethical Problems in the Delivery of Health Care, State of New Jersey, *Death and the Brain-Damaged Patient* (Trenton, NJ: State of New Jersey Commission on Legal and Ethical Problems in the Delivery of Health Care, 1992).
51. New Jersey Commission on Legal and Ethical Problems in the Delivery of Health Care, State of New Jersey, *The New Jersey Advance Directives for Health Care Act (and the Patient Self-Determination Act): A Guidebook for Health Care Professionals* (Trenton, NJ: State of New Jersey Commission on Legal and Ethical Problems in the Delivery of Health Care, 1992).
52. New York State Task Force on Life and the Law, *The Determination of Death* (New York, NY: New York State Task Force on Life and the Law, 1986).
53. New York State Task Force on Life and the Law, *Do Not Resuscitate Orders: The Proposed Legislation and Report of the New York State Task Force on Life and the Law* (New York, NY: New York State Task Force on Life and the Law, 1986).
54. New York State Task Force on Life and the Law, *The Required Request Law: Recommendations of the New York State Task Force on Life and the Law* (New York, NY: New York State Task Force on Life and the Law, 1986).
55. New York State Task Force on Life and the Law, *Life-Sustaining Treatment: Making Decisions and Appointing a Health Care Agent* (New York, NY: New York State Task Force on Life and the Law, 1987).
56. New York State Task Force on Life and the Law, *Fetal Extrauterine Survivability: Report of the Committee on Fetal Extrauterine Survivability to the New York State Task Force on Life and the Law* (New York, NY: New York State Task Force on Life and the Law, 1988).
57. New York State Task Force on Life and the Law, *Surrogate Parenting: Analysis and Recommenda-*

tions for Public Policy (New York, NY: New York State Task Force on Life and the Law, 1988).

58. New York State Task Force on Life and the Law, *Transplantation in New York State: The Procurement and Distribution of Organs and Tissues* (New York, NY: New York State Task Force on Life and the Law, 1988).
59. New York State Task Force on Life and the Law, *When Others Must Choose: Deciding for Patients Without Capacity* (New York, NY: New York State Task Force on Life and the Law, 1992).
60. Olick, R.S., Lowenstein, Sandler, Kohl, Fisher, and Boylan, Roseland, NJ, personal communication, March 1993.
61. Porter, J.P., "The Office for Protection From Research Risks," *Kennedy Institute of Ethics Journal* 2:279-284, 1992.
62. President's Commission for the Study of Ethical Problems in Medicine and Biomedical and Behavioral Research, *Defining Death* (Washington, DC: U.S. Government Printing Office, 1981).
63. President's Commission for the Study of Ethical Problems in Medicine and Biomedical and Behavioral Research, *Protecting Human Subjects* (Washington, DC: U.S. Government Printing Office, 1981).
64. President's Commission for the Study of Ethical Problems in Medicine and Biomedical and Behavioral Research, *Compensating Research Injury* (Washington, DC: U.S. Government Printing Office, 1982).
65. President's Commission for the Study of Ethical Problems in Medicine and Biomedical and Behavioral Research, *Making Health Care Decisions (with vols. 2 and 3 appendices)* (Washington, DC: U.S. Government Printing Office, 1982).
66. President's Commission for the Study of Ethical Problems in Medicine and Biomedical and Behavioral Research, *Splicing Life* (Washington, DC: U.S. Government Printing Office, 1982).
67. President's Commission for the Study of Ethical Problems in Medicine and Biomedical and Behavioral Research, *Whistleblowing in Biomedical Research* (Washington, DC: U.S. Government Printing Office, 1982).
68. President's Commission for the Study of Ethical Problems in Medicine and Biomedical and Behavioral Research, *Deciding to Forego Life-Sustaining Treatment* (Washington, DC: U.S. Government Printing Office, 1983).
69. President's Commission for the Study of Ethical Problems in Medicine and Biomedical and Behavioral Research, *Implementing Human Research Regulations* (Washington, DC: U.S. Government Printing Office, 1983).
70. President's Commission for the Study of Ethical Problems in Medicine and Biomedical and Behavioral Research, *Screening and Counseling for Genetic Conditions* (Washington, DC: U.S. Government Printing Office, 1983).
71. President's Commission for the Study of Ethical Problems in Medicine and Biomedical and Behavioral Research, *Securing Access to Health Care* (Washington, DC: U.S. Government Printing Office, 1983).
72. President's Commission for the Study of Ethical Problems in Medicine and Biomedical and Behavioral Research, *Summing Up* (Washington, DC: U.S. Government Printing Office, 1983).
73. Reich, W.T., Georgetown University, Washington, DC, "How Bioethics Got Its Name," remarks at the conference, "The Birth of Bioethics," Seattle, WA, September 1992.
74. Rothman, D.J., *Strangers at the Bedside: A History of How Law and Bioethics Transformed Medical Decisionmaking* (New York, NY: Basic Books, Inc., 1991).
75. Sullivan, L.W., Secretary, Department of Health and Human Services, statement, Nov. 2, 1989.
76. Sun, M., "New Bioethics Panel Under Consideration," *Science* 222:34-35, 1983.
77. U.S. Congress, House of Representatives, Committee on Government Operations, *Infertility in America: Why is the Federal Government Ignoring A Major Health Problem*, H. Rpt. 101-389 (Washington, DC: U.S. Government Printing Office, 1989).
78. U.S. Congress, House of Representatives, Committee on Government Operations, *Designing Genetic Information Policy: The Need for an Independent Policy Review of the Ethical, Legal, and Social Implications of the Human Genome Project*, H. Rpt. 102-478 (Washington, DC: U.S. Government Printing Office, 1992).
79. U.S. Congress, House of Representatives, Committee on Government Operations, Subcommittee

on Human Resources and Intergovernmental Relations, ''Medical and Social Choices for Infertile Couples and the Federal Role in Prevention and Treatment,'' hearing, July 14, 1988 (Washington, DC: U.S. Government Printing Office, 1989).

80. U.S. Congress, Office of Technology Assessment, *Human Gene Therapy—A Background Paper*, OTA-BP-BA-32 (Washington, DC: U.S. Government Printing Office, December 1984).
81. U.S. Congress, Office of Technology Assessment, *New Developments in Biotechnology: Ownership of Human Tissues and Cells*, OTA-BA-337 (Washington, DC: U.S. Government Printing Office, March 1987).
82. U.S. Congress, Office of Technology Assessment, *Life-Sustaining Technologies and the Elderly*, OTA-BA-306 (Washington, DC: U.S. Government Printing Office, July 1987).
83. U.S. Congress, Office of Technology Assessment, *Mapping Our Genes—The Human Genome Projects: How Big, How Fast?*, OTA-BA-373 (Washington, DC: U.S. Government Printing Office, April 1988).
84. U.S. Congress, Office of Technology Assessment, *Infertility: Medical and Social Choices*, OTA-BA-358 (Washington, DC: U.S. Government Printing Office, May 1988).
85. U.S. Congress, Office of Technology Assessment, *New Developments in Biotechnology: Patenting Life*, OTA-BA-370 (Washington, DC: U.S. Government Printing Office, April 1989).
86. U.S. Congress, Office of Technology Assessment, *Neural Grafting: Repairing the Brain and Spinal Cord*, OTA-BA-462 (Washington, DC: U.S. Government Printing Office, September 1990).
87. U.S. Congress, Office of Technology Assessment, *Genetic Monitoring and Screening in the Workplace*, OTA-BA-455 (Washington, DC: U.S. Government Printing Office, October 1990).
88. U.S. Congress, Office of Technology Assessment, *Cystic Fibrosis and DNA Tests: Implications of Carrier Screening*, OTA-BA-532 (Washington, DC: U.S. Government Printing Office, August 1992).
89. U.S. Congress, Office of Technology Assessment, *The Biology of Mental Illness*, OTA-BA-538 (Washington, DC: U.S. Government Printing Office, September 1992).
90. U.S. Congress, Senate, Committee on Government Operations, Subcommittee on Government Research, *National Commission on Health Science and Society*, hearings Mar. 7-8, 21-22, 27-28; Apr. 2, 1968 (Washington, DC: U.S. Government Printing Office, 1968).
91. U.S. Department of Health, Education, and Welfare, Ethics Advisory Board, *Report and Conclusions: HEW Support of Research Involving Human In Vitro Fertilization and Embryo Transfer* (Washington, DC: U.S. Department of Health, Education, and Welfare, 1979).
92. U.S. Department of Health, Education, and Welfare, Ethics Advisory Board, *Report and Recommendations: Research Involving Fetoscopy* (Washington, DC: U.S. Department of Health, Education, and Welfare, 1979).
93. U.S. Department of Health, Education, and Welfare, National Commission for the Protection of Human Subjects of Biomedical and Behavioral Research, *Research on the Fetus: Report and Recommendations*, DHEW Pub. No. (OS)76-127 (Washington, DC: U.S. Government Printing Office, 1975).
94. U.S. Department of Health, Education, and Welfare, National Commission for the Protection of Human Subjects of Biomedical and Behavioral Research, *Research Involving Prisoners* (Washington, DC: U.S. Government Printing Office, 1976).
95. U.S. Department of Health, Education, and Welfare, National Commission for the Protection of Human Subjects of Biomedical and Behavioral Research, *Disclosure of Research Information Under the Freedom of Information Act* (Washington, DC: U.S. Government Printing Office, 1977).
96. U.S. Department of Health, Education, and Welfare, National Commission for the Protection of Human Subjects of Biomedical and Behavioral Research, *Psychosurgery* (Washington, DC: U.S. Government Printing Office, 1977).
97. U.S. Department of Health, Education, and Welfare, National Commission for the Protection of Human Subjects of Biomedical and Behavioral Research, *Research Involving Children* (Washington, DC: U.S. Government Printing Office, 1977).

98. U.S. Department of Health, Education, and Welfare, National Commission for the Protection of Human Subjects of Biomedical and Behavioral Research, *The Belmont Report: Ethical Principles and Guidelines for the Protection of Human Subjects of Research* (Washington, DC: U.S. Government Printing Office, 1978).
99. U.S. Department of Health, Education, and Welfare, National Commission for the Protection of Human Subjects of Biomedical and Behavioral Research, *Ethical Guidelines for the Delivery of Health Services by DHEW* (Washington, DC: U.S. Government Printing Office, 1978).
100. U.S. Department of Health, Education, and Welfare, National Commission for the Protection of Human Subjects of Biomedical and Behavioral Research, *Institutional Review Boards* (Washington, DC: U.S. Government Printing Office, 1978).
101. U.S. Department of Health, Education, and Welfare, National Commission for the Protection of Human Subjects of Biomedical and Behavioral Research, *Research Involving Those Institutionalized as Mentally Infirm* (Washington, DC: U.S. Government Printing Office, 1978).
102. U.S. Department of Health, Education, and Welfare, National Commission for the Protection of Human Subjects of Biomedical and Behavioral Research, *Special Study: Implications of Advances in Biomedical and Behavioral Research* (Washington, DC: U.S. Government Printing Office, 1978).
103. U.S. Department of Health and Human Services, Ethics Advisory Board, *The Request of the Centers for Disease Control for a Limited Exemption From the Freedom of Information Act* (Washington, DC: U.S. Department of Health and Human Services, 1980).
104. U.S. Department of Health and Human Services, Ethics Advisory Board, *The Request of the National Institutes of Health for a Limited Exemption From the Freedom of Information Act* (Washington, DC: U.S. Department of Health and Human Services, 1980).
105. U.S. Department of Health and Human Services, National Institutes of Health, Human Fetal Tissue Transplantation Research Panel, *Report of the Human Fetal Tissue Transplantation Research Panel, Volume I* (Bethesda, MD: National Institutes of Health, 1988).
106. U.S. Department of Health and Human Services, National Institutes of Health, Human Fetal Tissue Transplantation Research Panel, *Report of the Human Fetal Tissue Transplantation Research Panel, Volume II* (Bethesda, MD: National Institutes of Health, 1988).
107. Wallace-Brodeur, P.H., "Community Values in Vermont Health Planning," *Hastings Center Report* 20(5):18-19, 1990.
108. Weisbard, A.J., "The Role of Philosophers in the Public Policy Process: A View From the President's Commission," *Ethics* 97:776-785, 1987.
109. Wikler, D., "Symposium on the Role of Philosophers in the Development of Public Policy: Introduction," *Ethics* 97:775, 1987.
110. Wyngaarden, J.B., Director, National Institutes of Health, memorandum to Robert E. Windom, assistant secretary for Health, U.S. Department of Health and Human Services, Jan. 19, 1989.
111. Yesley, M.S., "The Use of an Advisory Commission," *Southern California Law Review* 51:1451-1469, 1978.
112. Zurer, P.S., "Panel Recommendations Would Hobble Scientific Misconduct Investigations," *Chemical and Engineering News* 70(44):19-20, 1992.

Prospects for the Future 3

Over the past two decades, the creation of Federal bioethics commissions resulted from a desire for mechanisms to articulate common values and foster consensus about biomedical advances in the face of cultural and religious heterogeneity. With the accelerated pace of technological innovation, it will become increasingly important for policymakers to understand the bioethical issues of such advances. **How best to incorporate bioethical analyses into policy decisionmaking is a challenge facing Congress today** (8,26-29,40,42). If Congress decides to create a new Federal bioethics body, what type of effort should it consider based on past experience? Which particular factors promote success, and which should be avoided?

WHAT ROLE COULD A COMMISSION PLAY?

The National Commission for the Protection of Human Subjects of Biomedical and Behavioral Research (National Commission), the Ethics Advisory Board (EAB), the President's Commission for the Study of Ethical Problems in Medicine and Biomedical and Behavioral Research (President's Commission), and the Biomedical Ethics Advisory Committee (BEAC) were all Federal responses to address ethical disputes in medical practice and the conduct of biomedical research. With the conclusion or demise of each of these efforts came calls for a new Federal entity (1,12,15,18,20,34,48). Today, however, no public body exists for the exclusive deliberation of complex bioethical dilemmas. **For nearly 4 years—the longest period of time since bioethics burgeoned as a discipline—the Federal Government has been without a formal forum that addresses bioethical issues. In fact, a fully operational body has not existed in over a decade.**

The current void has not gone unnoticed by either the professional communities or policymakers. Today, both parties increasingly decry the lapse (26,28,83), just as Members of Congress (70) and experts (1) sought a new venue for bioethics after the President's Commission concluded its work. A sense of urgency permeates current appeals for two reasons: the accelerated pace of biomedical innovation and the length of time that has elapsed since the last government initiative.

Even without a formal Federal effort, however, bioethical analyses have been incorporated into selected public policy analyses. Several OTA reports include bioethical analyses (73-82), and at the request of Congress or the executive branch, the Institute of Medicine (IOM)[1] has addressed ethical issues in relevant reports (25,31,33,35, 50,58,71). As described in chapter 2, the Ethical, Legal, and Social Issues programs of the National Institutes of Health (NIH) and U.S. Department of Energy currently fund bioethics research related to the Human Genome Project.

Today, policy decisionmakers find themselves besieged with bioethical issues seeking resolution (8,26-29,40,42). The intellectual fecundity of these issues is apparent by the presence of a growing corps of bioethicists (65) and bioethics organizations. The American Association of Bioethics was launched in March 1993 (5,14), and the International Association of Bioethics held its inaugural congress in 1992. A few State efforts have succeeded in exploring bioethical issues (ch. 2). In addition, many academic or private efforts exist. Several medical or research organizations have formed ethics groups, including the American Medical Association, American College of Obstetricians and Gynecologists, American Fertility Society, American Society of Human Genetics, American Public Health Association, and American Academy of Pediatrics. And while nongovernmental groups (e.g., the National Advisory Board on Ethics in Reproduction) lack the public sanction and authority that inhere with government-appointed bodies (88), nongovernmental associations play an important role in shaping the bioethics debate and have been particularly effective in framing the dialogue in countries without national bodies (43,64,67,88,90). Still, no Federal initiative with sufficient authority or visibility exists to systematically analyze the ethical implications of important issues such as genetic privacy, embryo and fetal research, and research involving people with mental illness.

Despite the lack of a Federal forum for bioethics, the development of different approaches involving many voices at many levels is viewed as beneficial by some, but insufficient by others. A widespread, pluralistic approach has advantages over a single national commission by fostering diversity; no issue becomes captive to any central authority. Another advantage is that individuals who will be the actual implementors or enforcers of the guidelines have more opportunities to participate in the process. Regional or local approaches also allow a community's values to be integrated into local political processes.

Nevertheless, the diversity in bioethics organizations—while bringing the debate to the State, local, or institutional level—cannot always succeed in addressing areas that require expansive access to information and expertise. As much as a practitioner or local organization would like to keep abreast of developments in bioethical analyses, expedient decisions often must be made. The availability of guidance that is consistent with a broader, national approach can be invaluable and sometimes even preferred (91)—i.e., ad hoc decisions might be appropriate

[1] IOM is part of the National Academy of Sciences, a private, nonprofit organization established by Federal charter to advise the Government on scientific matters. The majority of its studies are undertaken at the request of the executive branch. In 1992, it began, on its own initiative, a broad study of methods of bioethical problem solving by society, including government, community bodies, professional societies, and religious groups. The study is not confined to formal commissions and agencies, and publication is expected in early 1994.

sometimes, but guidance in the form of generally agreed upon principles helps maintain a level of consistency and comparability across the health and legal professions (60). A Federal body can identify areas of national consensus or division.

Thus, while OTA uncovered a range of opinions on the optimal framework to incorporate bioethics into U.S. public policy decisionmaking, OTA found strong sentiment on the need for a Federal initiative, or initiatives, that would involve diversely trained individuals to monitor, analyze, and report on the interface between ethics and medicine, health care, and biomedical and behavioral research. Such organizations could be charged with the responsibility of informing legislators, regulators, judges, health care providers, scientists, and the lay public about the ethical implications raised by new situations in medicine and biomedical research.

There are pitfalls attendant to centralization, however, including a tendency to lose flexibility in interpretation, diffusion arising from forced consensus (6,41,89), and the potential for capture by political interests. Yet, centralization brings authority to a process that is rarely achieved with decentralization. A Federal effort generally can exceed private or State resources. It also carries cachet, as well as a nationwide power to gather data, convene meetings, generate relevant analyses, and invite testimony. The process by which this background paper was produced demonstrates this value: Though allocated a limited budget and a short timeframe, the cooperation of workshop participants, survey respondents, interviewees, and reviewers moved forward because it was a congressional effort. For better or worse, professionals, Members of Congress, or State legislators assign different weight to groups that they have created than to other groups in operation. Nevertheless, private, professional, local, and national approaches are not mutually exclusive; all are desirable.

Government-sanctioned commissions allow debates about contentious issues to go forward in a somewhat less politicized way than is possible on the floors of Congress or a State legislature. National commissions provide a vehicle to handle issues that are amenable to consensus building, or at least to an elaboration of conflicting views. Ideally, they garner the esteem of policymakers and experts by serving as a forum to:

- crystallize a consensus or delineate points of disagreement;
- identify emerging issues;
- defuse controversy or delay decisionmaking;
- propose regulations, develop guidelines, or formulate policy options;
- review implementation of existing law and policies;
- aid judicial decisionmaking;
- educate professionals and the public; and
- promote interdisciplinary research (15,18, 53,91).

Still, commissions tend not to be ground breaking intellectually (13), although they can summarize current thinking into a form meaningful for policymakers (13,15). Further, commissions can clarify issues and offer useful critiques of public policy, but they lack the moral and political authority to decide what ought to be done (57).

Nevertheless, the process of convening a commission for the Federal Government can be an opportunity to create the environment in which political action becomes possible by gathering policy relevant information and injecting it directly into the policy matrix. In doing so, commissions can often consider controversial issues independent of the regular political process and its constraints (72). Commissions cannot quiet all bioethical concerns, but can provide a broadly accepted basis for understanding the issues and propose particular policies to cover most situations (15).

The history of regulations governing the participation of humans in research provides an example of the validating powers of national commissions. In 1973, the then Department of Health, Education, and Welfare (DHEW) proposed rules on research involving human fetuses. The rules, which were published in revised form in 1974, generated a storm of controversy and resulted in heated debate in Congress. The bureaucrats who had prepared the rules had done a reasonable job of examining the literature and putting together thoughtful, well-articulated proposals, and they were surprised by the ensuing debate (52). Eventually, Congress created the National Commission and assigned it the task of addressing fetal research, which it did in about the required 4 months (85). What the Commission recommended was similar to that originally proposed, but this time the substance was received with praise and approval from all sides (52). In politically sensitive areas of debate, sometimes the messenger is more important than the message.

WHAT TYPES OF FORUMS SHOULD BE CONSIDERED?

Before considering the specific elements of any future effort, decisions about proposed structure and function must be addressed. Past Federal bioethics initiatives provide a guide should Congress decide to launch a new effort or efforts. **OTA identified three basic types of organizational models that should be considered:**

- **continuous/standing,**
- **term limited, and**
- **ad hoc.**

Do certain topics or areas of inquiry lend themselves to a particular structure? If so, then **the scope and issues Congress believes a commission must address could drive the type of policy body that would be most appropriate to establish.** Linking the class of issue to the commission structure might help ensure optimal consideration in a timely, effective, and economic manner. Less than this could endanger patients or research subjects, delay decisionmaking and lead to gaps in policy implementation, or interfere with vital research.

OTA identified two general classes of issues for which bioethical analyses have been applied: specific classes of, or protocols in, biomedical or behavioral research involving human subjects (table 3-1) **versus broad-based issues related to medical practices, health care, or the social implications of research** (table 3-2). The following sections analyze these two categories within the three general models: an ongoing body, a term-limited commission, and an ad hoc effort.

Standing Bodies

As part of its work, the National Commission concluded that ethical deliberations involving the review of protocols—or classes of protocols—arising from controversial biomedical and behavioral research (table 3-1) warranted a standing body. Despite its demise, EAB was chartered for this purpose and was the only Federal initiative intended as a continuing body. **OTA concurs with the National Commission's recommendation that a standing body is appropriate to consider the ethical implications of certain protocols or classes of federally funded biomedical and behavioral research. Research-related issues—e.g., AIDS vaccine protocols or clinical trials using human growth hormone in children without identifiable disease (2,69)—are ongoing, and any single proposal that raises novel ethical questions can appear suddenly.**

An ongoing entity in the model of EAB would be beneficial. Without an EAB-like body to provide guidance, the Federal Government has turned to separate ad hoc panels to perform precisely the functions of an EAB—including the Fetal Tissue Panel and the committee to examine human growth hormone trials in children more recently, as well as groups to evaluate other protocols in children or prisoners in the past. In fact, recognizing new questions can arise with any proposal, NIH recently established standing bodies called protocol implementation review committees to identify potentially sensitive intramural research (68,69); the new panels do not, however, examine extramural research.

Table 3-1—Biomedical Research Topics That May Raise Unresolved Ethical Issues

Topic
Clinical trials for anti-addictive medications
Clinical trials in children of synthetic human growth hormone for cosmetic versus therapeutic uses
Clinical trials in women and minorities
Compassionate uses of gene therapy outside controlled clinical trials
Conduct of AIDS vaccine trials
Drug trials and clinical studies of individuals with dementia
Drug trials and clinical studies of individuals with mental illness, e.g., schizophrenia
Embryo research
Fetal research
Genome research on aboriginal human populations
Involvement of women of childbearing age in drug trials
Large family pedigree genetics research
Research involving RU 486
Rules governing research where patients pay for clinical research through service fees
Update of what constitutes "minimal risk," "innovative therapy," "experimental treatment," and other terms of art embedded in current regulations

SOURCE: Office of Technology Assessment, 1993.

Nevertheless, the current climate to eliminate nonmandated Federal advisory groups poses a significant barrier to reconstituting an EAB-like body; the Clinton Administration has mandated the termination of not less than one-third of bodies not required by statute (Executive Order 12838; 58 FR 8207). **In the face of significantly shrinking numbers of Federal advisory committees, the prospect of DHHS reviving a former body seems unlikely. Thus, Congress could require DHHS to establish an ongoing panel to evaluate ethical issues raised by federally funded biomedical and behavioral research.** A congressional mandate for an EAB per se is not necessary, but a directive to establish one and clarify its scope ultimately might be needed. Without legislation, the future of a standing body likely will be held in abeyance—despite the fact that a backlog of research-related issues exists (table 3-1) and that the ever-accelerating pace of biomedical research is sure to generate more. Ironically, though the United States no longer has an ongoing body of this nature, it is the most common type of effort in other countries.

■ Term-Limited Commissions

The National Commission, the President's Commission, and BEAC were all bodies whose terms were limited by Congress through sunset provisions. As mentioned, the National Commission examined issues primarily related to human subjects research. The President's Commission's focus was largely policy analysis of broad-based topics related to clinical practice, though it also examined research-related issues. BEAC issued no reports.

Thus, on the surface, commissions that have a fixed term appear appropriate for either class of issue. Nonetheless, as just discussed, issues raised by controversial research appear best suited to a standing body. **Term-limited entities could attend to the types of issues that were the focus of most reports produced by the President's Commission—broad-based topics arising from Federal activities or interest in medicine, health care, or research** (table 3-2). **Again, however, establishing such an entity probably would require congressional legislation because of the move to fewer Federal advisory bodies** (Executive Order 12838; 58 FR 8207).

Since the United States established the President's Commission, similar bodies have been created abroad—most notably in Denmark,

Table 3-2—Possible Bioethical Issues for a Broad-Based Effort

Issue
Animal patents
Dilemmas arising in emergency care situations
Euthanasia
Genetic privacy
Health care providers' equity interests and self-referral
Organ transplantation issues, including availability and xenografts
Patenting human tissues, cells, or DNA
Research collaboration and conflicts of interest

SOURCE: Office of Technology Assessment, 1993.

France, and the United Kingdom (app. A)—but all appear to be permanent bodies. Despite this, OTA found little sentiment for establishing a permanent, broad-based bioethics initiative modeled after international efforts or those in New York and New Jersey. Of concern is allowing a commission to become a self-perpetuating body in search of a mission.

■ Ad Hoc Initiatives

By its nature, an ad hoc initiative is confined to a single topic, and one may be convened at any time (box 3-A). In these respects, integrating bioethics into public policy decisionmaking in an ad hoc manner offers the advantage of flexibility. Another advantage is that members can be selected for expertise on the specific topic at hand. Similarly, convening several groups means more people have an opportunity to closely influence the process. And of course if no ad hoc efforts are convened, no money will be spent.

However, **OTA found consensus that ad hoc initiatives are the least desirable mechanism to address bioethical dilemmas.** Each individual initiative requires a certain critical energy, financial support, and personnel level before it becomes functional. Repeatedly starting committees or panels to review controversial protocols or practices when commonalities among topics exist—and could be analyzed by a single body—not only results in a less than optimal use of time,

Box 3-A—Bioethics and the National Institutes of Health Revitalization Act of 1993

Congress has pursued legislation to prevent the Secretary of the Department of Health and Human Services from making unilateral decisions that deny, based on ethical considerations, funding for peer-reviewed, approved projects. Two events prompted this legislation: the Secretary's moratorium on fetal tissue transplantation research and the decision to withhold funding of the adolescent sexual behavior survey (59). The National Institutes of Health Revitalization Act of 1993 (S. 1 and H.R. 4) contains language intended to prohibit the Secretary from withholding funds for research on "ethical grounds" unless he or she convenes an ethics advisory board, or "ethics board." No definition of ethical grounds is offered; what constitutes ethical grounds appears to be left to the Secretary.

Despite the confusing terminology created by using "Ethics Advisory Boards" [sic] out of historical context, the entities created by this Act would be wholly dissimilar to the original EAB. These bodies would be implemented only if the Secretary refused to fund a successfully peer-reviewed proposal. Thus, they are ad hoc in nature, and can be viewed as a subset of review panels because they are instituted only in the rare instances in which funding is withheld or withdrawn on ethical grounds. A wide range of research proposals that are funded—but might benefit from an EAB review—go unevaluated.

In the event a Secretary withholds funds, he or she must appoint an ethics board after considering nominations for 30 days; 180 days later, the body must submit a report to the Secretary and Congress. The legislation directs that an ethics board shall be composed of 14 to 20 individuals and shall include at least one attorney, one practicing physician, one ethicist, and one theologian. No fewer than one-third and no more than one-half shall be biomedical or behavioral scientists. The board would be staffed by the National Institutes of Health and would expire 30 days after it submits its report.

Should the majority of the ethics board recommend that the Secretary not withhold the monies for the research on ethical grounds, the research shall be funded unless the Secretary finds that the board's recommendations were "arbitrary and capricious."

SOURCE: Office of Technology Assessment, 1993, based on "S. 1," *Congressional Record* 139:S1806-S1808, 1993; and "HR.4," *Congressional Record* 139: H1184-H1185, 1993.

money, and people, but does the public, policymakers, and other interested parties a disservice.

Additionally, the initial learning and acculturation process of any group requires time and effort (23,52,92) and must be repeated each time an ad hoc panel is convened. Later reports of the President's Commission, for example, were considered better by its executive director because of this phenomenon (17). The work of the National Commission also improved substantially over time; its report on prisoners was much more polished and sophisticated than its preceding report on fetuses (48,49).

Continuity also can contribute to a body's credibility (53). Because ad hoc efforts cease to exist after a topic is addressed, any credibility gained with the passage of time or completion of projects is immediately lost. That is, a standing or fixed-term body, by doing a good job on one topic, garners the respect of a variety of constituencies and can transfer that credibility to its handling of new topics (52).

WHAT ELEMENTS ARE IMPORTANT?

The experiences of the National Commission, EAB, President's Commission, and BEAC illuminate a variety of key considerations for any future effort to create a national bioethics board or commission. **In devising a strategy for addressing bioethical issues in a national policy context, OTA found six factors predominate regardless of the type of body established:**

- budget, including staffing;
- the charge (i.e., mandate and flexibility to control the agenda);
- appointment process;
- bureaucratic location;
- target audience(s); and
- reporting and response requirements.

Absent from this list is politics; the very nature of creating a new entity subjects each of these factors to the pressures and whims inherent in the political process. An inadequate or ill-suited approach in any single area can undermine the successful implementation of a new national commission or board, and a deficiency in a single aspect—e.g., funding or the appointment process—can doom an effort to total failure.

■ Funding

Although each factor is important, funding is foremost. A successful initiative begins with adequate funding (box 3-B). Sufficient funds to hire an adequate number of qualified, professional staff are essential; otherwise the entity is staffed piecemeal or by castoffs. Also necessary are monies for contract papers and public hearings or meetings beyond the Washington, DC locus. Commissions are most successful when they can weigh both empirical information and conceptual analyses to derive useful policy options. Thus, they must have the capability to gather data. Public meetings and hearings—providing an educational function for both commissioners and the public—are imperative. Finally, providing funds for publication and dissemination of a commission's work is essential.

The National Commission and the President's Commission had reasonable funds for staffing and activities—as initially did BEAC. Likewise, the commissions and their staffs had broad contracting authority and funds to bring in additional expertise, as well as to hear witnesses. Both capabilities served the staff well by expanding their knowledge base; staffing can also provide balance. For example, staff can weigh and analyze the persuasiveness of particular points of view without regard for the sophistication or eloquence of the defender of that position. Without staff to mediate between the more and less sophisticated presentations, a commission's policy documents run the risk of becoming skewed (4,8,14,17).

The National Commission did not have a specific appropriation, but during its 4-year lifespan received about $5 million from DHEW (93). Nevertheless, the need for specific appropri-

Box 3-B—What Might a New Effort Cost?

Estimating a potential budget for a new bioethics entity depends on many factors, including the number of members, staff, meetings, and reports. Start-up capital costs also will vary based on size and whether a new body is housed within an existing Federal agency or office, or created as an independent body. Despite these caveats, funding projections are possible. Budgets can be constructed by examining current Federal commissions and agencies performing policy analysis. Examining the expenditures of previous bioethics bodies is less useful because of the significant changes in basic business costs since their existence—e.g., rent, computer hardware and software, telecommunication technologies, travel for commission members, and salaries.

Although a comprehensive survey of budgets for commissions, task forces, or committees was not possible, funding levels for a new standing body (like the Ethics Advisory Board (EAB)) and a new term-limited commission (like the President's Commission) are presented based on OTA's experience with producing policy reports and a brief inspection of a few other efforts. For each, certain assumptions are described. Clearly, changes in any assumption would affect the funding required.

Standing Body: $744,000 annually. This budget assumes such a body will be located within an existing agency. It assumes 20 members will be appointed; by comparison, the charter proposed in 1988 for a reconstituted EAB envisioned 21 members (53 FR 35232). The cost also assumes five staff (i.e., five full time equivalents (FTEs) represented by four analytic and one support staff). Compared to the initial EAB's eight staff, this represents about a 40 percent reduction in FTEs. The cost also assumes contracting authority at 30 percent of staff costs. As with the proposed charter, it assumes ten 1-day meetings will be held—at least three outside of Washington, DC—and, thus, includes funds for travel and per diem, as necessary. The annual budget assumes three small documents (on the order of position papers, not full reports) and one full report per year, and includes printing and distribution costs. Overhead costs of computer literature searches, books, and journal subscriptions are included. However, because this body is assumed to be housed within an existing Federal agency or office, *the budget does not include costs such as rent or utilities. Nor are costs associated with personnel who manage procurement, reimbursement, computer support or other general services included.*

Term-limited Body: $1,920,000 per year for 4 years. This estimate assumes a new term-limited body will operate as an independent commission. The budget assumes 15 members will be appointed—4 more than the President's Commission. It also assumes 12 FTEs (9 professional and 3 support positions) will staff the new effort—approximately 40 percent below that of the President's Commission. The cost includes six 2-day meetings in Washington, DC, and four 1-day field hearings outside Washington, DC, annually. Also included are initial outlays for equipment and annual costs for rent, utilities, and other items (e.g., supplies, books, journals) and services (e.g., mail, photocopying, and computer support). The budget also assumes six reports will be produced over the 4-year term, and includes contracting authority—for papers and editing—equivalent to 40 percent of staff costs, as well as printing and distribution costs. In the first year, funds will be needed for certain capital costs (e.g., computers, phones, facsimile machines, and photocopiers) that will not be needed in subsequent years. However, contracting, meeting, and publication costs will be incurred chiefly in later years as the commission begins its formal work. OTA's calculations revealed the average annual cost would be about the same between the first and later years, despite the shift in spending priorities.

SOURCE: Office of Technology Assessment, 1993.

ations can be critical; EAB's experience demonstrates this point. EAB was born from the National Commission, which had anticipated that EAB would be a continuing body. However, in the course of congressional deliberations after the National Commission's conclusion, the President's Commission was established with funds initially diverted from the EAB because policymakers failed to distinguish their distinct purposes (17,18). On the other hand, fiscal support could have been drawn from other U.S Department of Health and Human Services (DHHS) funds after 1980—as presumably was intended when plans to recharter EAB emerged in 1988 (53 FR 35232).

■ Mandate and Agenda Setting

The focus and mandate—global issues versus research issues, regulatory or advisory—influence an initiative's authority or lack thereof. Additionally, as discussed earlier in this chapter, the form a bioethics body takes—standing versus limited term—can be linked to its agenda and mandate. Thus, delineating a mechanism to set the scope of topics also is a consideration for Congress.

A commission's charge should be structured to provide guidance, if not requirements, for the selection of topics or issues for study. Circumscribing too narrow a function obviates the potential early warning benefit of bioethics commissions. Drawing too broad a boundary could move a commission to examine issues—e.g., health care reform—that Congress or the President have tasked to others.

Since Federal regulation over medical practices or biomedical research is limited to federally funded activities, Congress might restrict an agenda to areas with clear Federal authority—e.g., patenting animals. On the other hand, an agenda need not be limited by current Federal regulatory authority. Issues that are currently entirely matters of State law—e.g., assisted suicide (32)—could fall under the Federal rubric should they someday involve Federal reimbursement policies. Moreover, government deliberations can provide important information and identify common ground in areas generally the domain of States—e.g., animal-to-human organ transplants or assisted suicide—even when no Federal regulatory or funding role exists.

Each of the three previous congressionally created bioethics bodies was required to assess certain specific topics, but their flexibility to embark on analyses of other issues varied. A **combination of mandated studies and the opportunity for commissioners and staff to identify emerging issues maximizes the use of talent, time, and money. No single best mechanism, however, clearly prevails.**

One approach would be for the commission itself—rather than the sponsor—to initiate studies of nonmandated topics, under the belief that members and staff are in the best position to select study topics because of their expertise and detachment from political structures. That is, if properly and appropriately appointed, a commission should embody the capacity to select topics that are of shared concern to all (21). On the other hand, federally funded commissions are public bodies created by officials who will be held accountable to the public. To address this conflict, Congress or the President could be permitted to suggest or mandate topics—with the Commission retaining, or not retaining, the authority to refuse such suggestions.

Congress struck a balance in the authority it gave the President's Commission. It required that specific topics be addressed, but also provided an elastic clause that allowed either the Commission or the President to add additional topics. To prevent the use of the President's Commission for special interests, Congress explicitly excluded itself from being able to provide further topics; a practice, however, that would seem unlikely to work for a standing body. In fact, the President's Commission responded to a single additional request by President Carter and his Science Advisor and explored the ethics of gene therapy

(17). Similarly, the President's Commission undertook the issue of foregoing life-sustaining treatment on its own volition, and the report became one of its most influential and lasting (10,17,18,37). The National Commission also initiated topics on its own (49).

Finally, depoliticizing a commission's mandate is impossible (6), but a priori avoiding extremely politicized issues, such as abortion, enables a commission to be more efficient and productive (as in the case of the New York State Task Force (56)). The short life of BEAC—caught in the abortion controversy crossfire—provides the most persuasive argument for this approach (19).

■ Appointment Process and Composition

A commission's mandate affects the nominating process. In turn, commission (and staff) structure and composition shape substance (62). With a narrow scope and small size, the tendency might be to select individuals with narrow, predetermined ideologic views for membership, thus diminishing the chances for success. On the other hand, a mandate created for specific topical needs, as for the National Commission, New Jersey Bioethics Commission, and New York State Task Force, can be executed effectively if the body's membership is balanced to reflect diversity and specific areas of expertise.

Today's explosive growth in the field of bioethics and in the number of individuals described as bioethicists has enhanced the Nation's capacity to discuss bioethical issues. At the same time, these developments increase the risk of bodies rendering commentary that is ingrown or out of touch with the real worlds of health care providers, scientists, or the lay public. Thus, any new organization must be representative of society at large.

Diversity in race, ethnicity, gender, and professional experience is a paramount factor in appointing commissioners and staff (17,44,66). **Ethics involves values, and a commission with monolithic membership or staffing cannot hope to adequately represent the diverse range of perspectives in American society.** Additionally, the processes of a bioethics commission—that of formulating guidelines or regulations—is part social knowledge (appreciation of the problems rooted in experience) and part theory (ethical, legal, and philosophical). This means a deliberative body should be comprised of both practitioners and theoreticians.

The National Commission's success, for example, has been credited to its multidisciplinary and multiexperiential composition (44,66), although whether it was stacked in favor of research has been debated (3,38,61). Similarly, the success of the President's Commission has been attributed to the broad range of professionals that comprised it (17,62), although whether philosophy was adequately integrated has been questioned (11); on the other hand, questions also have been raised about the role of philosophers in addressing public policy issues (7,39,47,54,55).

Ideology is a destructive criterion in appointing a bioethics committee. While selecting members solely on the basis of their stance on a particular issue—such as abortion—might be viewed by special interests as useful, such an approach is shortsighted and likely to create gridlock. There is no way of predicting which way people will move on issues with which they are unfamiliar. Focusing solely on the views of potential panelists for one contentious issue, such as abortion, can delay the actual work of a committee, perhaps indefinitely; other issues, for which there might be broad consensus, are not given the floor.

The experience with BEAC illustrates this point. When BEAC was formed, it was believed that, although the membership was deliberately and strategically split on abortion, the panel would split in many different ways once it addressed more general issues—such as genetics. This supposition was apparent by the second meeting. It never got the chance to develop, however, because the rancor over abortion—and congressional focus on the prochoice/antiabortion makeup of BEAC members—prevented the congressional Biomedical Ethics Board (BEB), the oversight body comprised of Senators and Representatives, from permitting further work to proceed (18,20).

OTA found no consensus on the optimal size of a commission. A smaller body, e.g., the National Commission with 11 members, lends itself to quicker development of the interpersonal dynamics and collegiality necessary for consensus building; it is also easier to handle administratively, but might bog down in ideological demagoguery. In contrast, a larger body, such as the current 41-member French commission, could be more inclusive and might be less susceptible to political line drawing (46). On the other hand, a body this large might be unwieldy and require subcommittees or task forces to work effectively, which adds an additional layer of bureaucracy.

OTA also found no consensus on the ideal mechanism to appoint members. Successful commissions have been appointed wholly by the President, his agency designee, or by a statutorily prescribed formula that usually involves the President, the Speaker of the House, the House Minority Leader, and the Senate Majority and Minority Leaders (72). Another model is the Advisory Panel for Alzheimer's Disease. This independent panel is congressionally mandated to advise DHHS and is staffed by the National Institute on Aging, but OTA appoints panel members—a process viewed as both impartial and expeditious (20). OTA also successfully appoints members to the Prospective Payment Assessment Commission and the Physician Payment Review Commission. BEAC's members were appointed solely by its congressional oversight board, the BEB, and BEAC expired due to BEB's infighting when a vacancy needed filling.

OTA did find general consensus on the merits of rotating membership. With the National Commission, members were appointed and toiled for the full 4 years of the commission's life. One former member of the National Commission commented that by the conclusion of its work, ideas had become less fresh and responses reflexive, rather than open and discursive (44). Infusing new ideas and personalities by limiting members' terms carries some cost, however. Commission dynamics take time to develop; overly short term lengths would strain a consensus building process. The membership of the President's Commission was rotated and, on occasion, rotations disrupted the process of completing reports in progress (15,17). Nevertheless, rotating membership keeps a body from appearing to be politically locked in and should be favored.

■ Location

Deciding where to locate a Federal bioethics organization depends on its mission and purpose. In some cases it might be best to locate an ethics advisory group in the agency whose work it reviews, thereby maximizing the chance that its recommendations will be implemented. In other cases, the work of the advisory body might be too closely related to the work of the agency to accomplish anything—i.e., an independent body would be optimal.

The National Commission, for example, was formed because of growing concern about the treatment of human subjects in research. A crucible of debate over fetal research forced the issue, resulting in policy discussions on research involving human subjects. Most of these issues fell squarely in the domain of DHEW/DHHS, and Congress placed the National Commission within that agency in 1974. In retrospect, although the National Commission operated with a good deal of autonomy (63), one factor that contributed to the Commission's success was its close proximity to the agency it was reviewing and to which it made its recommendations (23,52). On the other hand, locating a body within the concerned agency could create a perception of conflict—i.e., that the ethics body is not wholly independent. EAB was housed within DHEW/DHHS, and although the impact of this body was less than that of the National Commission, it was significant (51). Had it been given time, its effectiveness in reviewing the Department's work could have been fully evaluated: Either its work would have been successful in terms of departmental implementation or it would have been embedded in the bureaucracy and ignored (52).

In contrast, the President's Commission has been deemed a success, in part due to its independent standing and freedom from Congress and the executive branch. Had it been integrated within DHHS, it is uncertain whether the Commission would have survived political interference by Congress and the executive branch, given the political climate of the early 1980s (17).

A prime example of location undermining a body's success was the failed 1985 attempt to establish BEB and BEAC. BEB created a partisan logjam along abortion lines that prevented the deliberative committee—BEAC—from ever completely considering an issue, and BEAC's demise can be attributed directly to congressional wrangling and lack of independence (18-20).

Besides these models—within an agency of the executive branch, an independent commission of the Federal Government, and attached to Congress—proposals also have suggested that OTA or IOM house a bioethics body (1,70). When such proposals were under consideration 8 years ago, OTA, as a congressional agency, was perceived as potentially subject to excessive politicization (1). IOM was criticized as a body ultimately captured by medical interests (1), even though IOM policy and membership includes nonmedical perspectives. The passage of time and the increased experience and prominence of both OTA and IOM might now mute these objections. Notably, IOM is not subject to the Federal Advisory Committee Act (5 U.S.C. Ap. 2, § 1 et seq.), which requires open meetings. IOM routinely reserves the right to conduct closed committee meetings, which is seen as a strength (30) or a weakness (14,15,92). With the exception of the New York State Task Force, all Federal and State bioethics bodies that OTA examined did not hold closed deliberations, though many international entities do.

Thus, **in considering location, Congress can look to a range of historical models and proposals.** It could also examine whether the locus for a bioethics commission should be in the Executive Office of the President—e.g., attached to the Office of Science and Technology Policy (OSTP). OSTP, however, has limited familiarity with biomedical ethics issues, although the legislation establishing the President's Commission mandated that OSTP's director have a liaison with the Commission (42 USC § 300v). Furthermore, while locating a commission in OSTP might lend stature to the effort, Congress would need to address the historical understaffing of this office (24,45).

■ Client

Target audiences for the work of a new bioethics entity include Congress, the executive branch, the academic community, health care providers, and the public. As just discussed, a body's bureaucratic location might define the primary client, but any new effort will affect—and hence should strive to serve—all parties. For a bioethics entity to operate effectively, restraints and controls must be in place to discourage or prevent political meddling with the staff or the conditions for operation—regardless of where the body is housed or who the principal client is. **OTA found consensus on the need for autonomy and independence from both congressional interference and mischief from the executive branch.**

Through the 1970s, Members of Congress, the President, and executive branch agency personnel—regardless of political affiliation—largely viewed commissions from the perspective of patrons of a process, not players in it (23). This culture reversed in the 1980s, resulting in executive branch interference with Federal regulations—e.g., the lack of an EAB to review protocols for human in vitro fertilization (IVF) (ch. 2; 77). Also, if an EAB had been in place, the piecemeal funding and staffing of the NIH Fetal Tissue Transplantation Research Panel would have been unnecessary. Similarly, intrusion by Congress resulted in the failure of BEAC to ever initiate a project (17,18).

Finally, as mentioned in chapter 2, the lay public is increasingly interested in and involved with resolving bioethical issues. The ability to operate in relative obscurity in the early history of biomedical ethics contributed to the success of the National Commission (44,66,92). Such a situation would be impossible today; intense public scrutiny exacerbated the difficulties experienced by BEAC (16,20).

Reporting and Response Requirements

To whom and how the group shall report, in what manner, and what response should be required are key factors for congressional consideration. In fact, **to whom a group should report its final work seems to raise little controversy; what is problematic is injecting politics during the deliberative process.** In practice, a new commission could be required to report to Congress, the executive branch, or both. For example, a new body could be responsible primarily to the executive branch, with Congress maintaining its traditional oversight role.

Even the most successful attempts to tackle bioethical issues will be ineffective if the results of those deliberations are censored or poorly distributed. The National Commission published its own work, which was inadequately distributed, and EAB's work remains even more difficult to trace because of poor dissemination. In contrast, the President's Commission greatly improved the manner in which findings were reported and distributed; reports were published and sold by the U.S. Government Printing Office. Because one goal of bioethics commissions is public and professional education, adequate funds and a plan for widespread public dissemination and reporting beyond the designated client are vital.

Nevertheless, whether and how the client must respond is probably more important than how the commission must report and disseminate (23,52). Should Congress create a new commission or board, it could include a forcing clause for accountability of the target client(s). For example, EAB addressed such controversial issues as fetoscopy, research with the human embryo, and freedom of information and the early release of clinical trial data. Its recommendations, however, were largely ignored by its client, the executive branch—even when the report was unanimously approved (52). The most glaring example of this was the 1979 report on IVF (84). Yet, by using it as background for professional practice guidelines, the American Fertility Society and other organizations interested in conducting IVF research in a responsible and ethical manner implicitly endorsed this report.

In contrast, Congress included a "forcing clause" in the legislation that created the National Commission. The clause did not require the Secretary to accept every recommendation made, but it did require the Secretary to accept recommendations within 180 days or publish reasons for not accepting them in the *Federal Register*. Such a clause forces some sort of decision or action on the part of Federal officials in response to a report. In fact, EAB was created as a result of a National Commission recommendation. The Secretary had to accept the recommendation or publish reasons for not accepting it, and the political untenability of rejecting an EAB was greater than the risk of creating it (52). Ironically, an unanticipated outcome was Secretarial censorship of EAB activities because the regulations associated with chartering EAB, unlike the legislation creating the National Commission, forced no response to EAB reports (52). Even so, a forcing clause does not guarantee responsiveness: In violation of the law (48), DHHS has failed to issue final regulations related to the National Commission's work on research involving the "mentally infirm" (87).

Finally, an open question is whether a commission should be directed to achieve and report consensus on an issue or to consider and articulate the merits of competing values (1,41,89). Congress could look to three successful models: the National Commission and President's Commission for consensus and OTA for analyzing the range of viewpoints.

SUMMATION AND PROSPECTUS

Does the United States need a government-sanctioned body, or bodies, dedicated to deliberating about the ethical issues raised by biomedical research, medical innovation, and health care? What have past efforts accomplished?

In only two decades, the U.S. Government's forays into bioethics have had lasting, measurable impacts (15,22,36,48,62). Federal regulations to protect human research subjects owe their existence in their current form to the National Commission—e.g., the National Commission's report on research involving children (86) raised controversial issues, but the guidelines finally proposed received praise and approval from all sides (49). In clinical practice, for example, the President's Commission shaped subsequent public debate in health care settings, legislatures, and courts about patient directives on life-sustaining treatments.

If Congress decides to create a new bioethics commission, it can look to the history of Government's involvement in bioethics for a wealth of experience and information. Although it is difficult to generalize whether a particular factor is specifically associated with ultimate success or failure—each commission had a slightly different model or existed in a different political climate—lessons can be learned from the National Commission, EAB, President's Commission, and BEAC. By examining this rich history of activity, OTA found six specific elements contributed to the success or failure of past efforts and so should be considered in devising future strategies. Not surprisingly, the budget is important, but mandate, appointment process, bureaucratic location, targeted client, and reporting and response requirements are also key. Absent from the list is politics, since creating a new body is inherently political, and the system will affect each factor.

Whether a standing, term-limited, or ad hoc commission should be established might depend on the type of issues Congress would like analyzed. An ongoing body in the model of EAB appears superior for examining questions raised by controversial research involving human subjects. A term-limited body like the President's Commission can address both research-related issues and broad-based topics in bioethics, but might be best for the latter if a standing body is available to address research topics. Still, in an era of shrinking numbers of Federal advisory bodies, the barriers involved in creating two forums, though for distinct purposes, could prove insurmountable. Ad hoc commissions can handle both categories, but OTA found consensus that ad hoc panels are less desirable—i.e., to be favored only as a last resort.

Past Federal bioethics efforts have been varied, innovative, and largely successful, but not enduring. Today, Congress stands at a crossroad. How best to incorporate bioethical analyses into policy decisionmaking is the issue currently facing Congress—one made especially difficult as fiscal realities mean fewer Federal advisory bodies and fewer staff to support them. Congress must decide what opportunities to seize, and when and how to move forward. Regardless of whether it creates a new, broad-based commission, directs DHHS to establish an EAB with a new mandate, or both, Congress should somehow provide a voice for biomedical ethics in public policy.

CHAPTER 3 REFERENCES

1. Abram, M.B. and Wolf, S.M., "Public Involvement in Medical Ethics: A Model for Government Action," *New England Journal of Medicine* 310:627-632, 1984.
2. Anderson, C., "Group Asks NIH to Stop Growth Hormone Trials," *Nature* 358:4, 1992.
3. Annas, G.J., "Report on the National Commission: Good as Gold," *Medicolegal News* 8(6):4-7, 1990.
4. Asch, A., Boston University of School of Social Work, Boston, MA, remarks at "Biomedical Ethics in U.S. Public Policy," a workshop sponsored by the Office of Technology Assessment, U.S. Congress, Dec. 4, 1992.
5. *Associated Press*, "Bioethics Association Will Be Based in Salt Lake," Mar. 25, 1993.
6. Bayer, R., "Ethics, Politics, and Access to Health Care: A Critical Analysis of the President's Commission for the Study of Ethical Problems in Medicine and Biomedical and Behavioral Research," *Cardozo Law Review* 6:303-320, 1984.
7. Benjamin, M., "Philosophical Integrity and Policy Development in Bioethics," *Journal of Medicine and Philosophy* 15:375-389, 1990.
8. Biden, J.R., Jr., U.S. Senate, statement during debate on S. 1, National Institutes of Health Revitalization Act of 1993, *Congressional Record* 139:S1787-S1788, Feb. 18, 1993.
9. Brock, D.W., "Symposium on the Role of Philosophers in the Development of Public Policy: Truth or Consequences—The Role of Philosophers in Policymaking," *Ethics* 97:786-791, 1987.
10. Brock, D.W., Brown University, Providence, RI, personal communication, March 1993.
11. Brody, B.A., "The President's Commission: The Need to Be More Philosophical," *Journal of Medicine and Philosophy* 14:369-383, 1989.
12. Bunker, J.P. and Fowles, J., "The President's Commission: Do We Need a Sequel?," *Hastings Center Report* 13(5):10-11, 1983.
13. Callahan, D., The Hastings Center, Briarcliff Manor, NY, personal communication, March 1993.
14. Caplan, A.L., University of Minnesota, Minneapolis, MN, personal communication, March 1993.
15. Capron, A.M., "Looking Back at the President's Commission," *Hastings Center Report* 13(5):7-10, 1983.
16. Capron, A.M., "Bioethics on the Congressional Agenda," *Hastings Center Report* 19(2):22-23, 1989.
17. Capron, A.M., University of Southern California Law Center, Los Angeles, CA, remarks at "Biomedical Ethics in U.S. Public Policy," a workshop sponsored by the Office of Technology Assessment, U.S. Congress, Dec. 4, 1992.
18. Cook-Deegan, R.M., "Finding a Voice for Bioethics in Public Policy: Federal Initiatives in the United States, 1974-1991," unpublished manuscript, November 1992.
19. Cook-Deegan, R.M., Institute of Medicine, Washington, DC, remarks at "Biomedical Ethics in U.S. Public Policy," a workshop sponsored by the Office of Technology Assessment, U.S. Congress, Dec. 4, 1992.
20. Cook-Deegan, R.M., Institute of Medicine, Washington, DC, personal communication, March 1993.
21. Faden, R.R., Johns Hopkins School of Hygiene and Public Health, Baltimore, MD, remarks at "Biomedical Ethics in U.S. Public Policy," a workshop sponsored by the Office of Technology Assessment, U.S. Congress, Dec. 4, 1992.
22. Faden, R.R. and Beauchamp, T.L., *A History and Theory of Informed Consent* (New York, NY: Oxford University Press, 1986).
23. Fletcher, J.C., University of Virginia, Charlottesville, VA, remarks at "Biomedical Ethics in U.S. Public Policy," a workshop sponsored by the Office of Technology Assessment, U.S. Congress, Dec. 4, 1992.
24. Friedman, T.L., "Clinton Trimming Lower-Level Aides: Savings of $10 Million a Year Seen at the White House," *New York Times*, Feb. 11, 1993.
25. Hanna, K.E. (ed.), Committee to Study Biomedical Decisionmaking, Institute of Medicine, *Biomedical Politics* (Washington, DC: National Academy Press, 1991).
26. Hatfield, M.O., U.S. Senate, statement during debate on H.R. 2507, National Institutes of Health Reauthorization Act of 1991, *Congressional Record* 138:S4719-S4724, Apr. 2, 1992.
27. Hatfield, M.O., U.S. Senate, remarks at "Biomedical Ethics in U.S. Public Policy," a work-

shop sponsored by the Office of Technology Assessment, U.S. Congress, Dec. 4, 1992.

28. Hatfield, M.O., U.S. Senate, statement during debate on S. 1, National Institutes of Health Revitalization Act of 1993, *Congressional Record* 139:S1792-S1796, Feb. 18, 1993.
29. Helms, J., U.S. Senate, statement during debate on S. 1, National Institutes of Health Revitalization Act of 1993, *Congressional Record* 139:S1784-S1787, Feb. 18, 1993.
30. Henderson, D.A., Office of Science and Technology Policy, Washington, DC, personal communication, March 1993.
31. Hogness, J.R. and Van Antwerp, M. (eds.), Institute of Medicine, *The Artificial Heart: Prototypes, Policies, and Patients* (Washington, DC: National Academy Press, 1991).
32. Holder, A.R., Yale University School of Medicine, New Haven, CT, personal communication, March 1993.
33. Institute of Medicine, *The Elderly and Functional Dependency* (Washington, DC: National Academy Press, 1977).
34. Institute of Medicine, "Ethical Problems in Medicine and Biomedical and Behavioral Research," draft minutes of a meeting, June 17, 1983.
35. Institute of Medicine, *Medically Assisted Conception: An Agenda for Research* (Washington, DC: National Academy Press, 1989).
36. Jonsen, A.R., "Public Policy and Human Research," *Biomedical Ethics Reviews*, J.M. Humber and R.T. Almeder (eds.) (Clifton, New Jersey: Humana Press, 1984).
37. Jonsen, A.R., University of Washington, Seattle, WA, personal communication, March 1993.
38. Jonsen, A.R. and Yesley, M., "Rhetoric and Research Ethics: An Answer to Annas," *Medicolegal News* 8(6):8-14, 1980.
39. Kamm, F.M., "The Philosopher as Insider and Outsider," *Journal of Medicine and Philosophy* 14:347-374, 1990.
40. Kassebaum, N.L., U.S. Senate, statement during debate on S. 1, National Institutes of Health Revitalization Act of 1993, *Congressional Record* 139:S1784-S1787, Feb. 18, 1993.
41. Katz, J., "Limping Is No Sin: Reflections on *Making Health Care Decisions*," *Cardozo Law Review* 6:243-265, 1984.
42. Kennedy, E.M., U.S. Senate, statement during debate on S. 1, National Institutes of Health Revitalization Act of 1993, *Congressional Record* 139:S1784-S1787, S1793, Feb. 18, 1993.
43. Kimura, R., "Ethics Committees for 'High Tech' Innovations in Japan," *Journal of Medicine and Philosophy* 14:457-464, 1989.
44. King, P.A., Georgetown University Law Center, Washington, DC remarks at "Biomedical Ethics in U.S. Public Policy," a workshop sponsored by the Office of Technology Assessment, U.S. Congress, Dec. 4, 1992.
45. Knezo, G.J., "White House Office of Science and Technology Policy: An Analysis," *CRS Report for Congress* (Washington, DC: Congressional Research Service, 1989).
46. Knoppers, B.M., University of Montreal Faculty of Law, Montreal, Canada, remarks at "Biomedical Ethics in U.S. Public Policy," a workshop sponsored by the Office of Technology Assessment, U.S. Congress, Dec. 4, 1992.
47. Kymlicka, W., "Moral Philosophy and Public Policy: The Case of NRTs (new reproductive technologies)," *Bioethics* 7:1-26, 1993.
48. Levine, R.J., *Ethics and Regulation of Clinical Research* (2nd ed.) (Baltimore, MD: Urban and Schwarzenberg, 1986).
49. Levine, R.J., Yale School of Medicine, New Haven, CT, personal communication, March 1993.
50. Lonergan, E.T. (ed.), Committee on a National Research Agenda on Aging, Institute of Medicine, *Extending Life, Enhancing Life: A National Research Agenda on Aging* (Washington, DC: National Academy Press, 1991).
51. McCarthy, C.R., "Experience With Boards and Commissions Concerned With Research Ethics in the United States," *Progress in Clinical and Biological Research* 128:111-122, 1983.
52. McCarthy, C.R., Kennedy Institute of Ethics, Washington, DC, remarks at "Biomedical Ethics in U.S. Public Policy," a workshop sponsored by the Office of Technology Assessment, U.S. Congress, Dec. 4, 1992.
53. Mendeloff, J., "Politics and Bioethical Commissions: 'Muddling Through' and the 'Slippery Slope'," *Journal of Health Politics, Policy, and Law* 10:81-92, 1985.

54. Menzel, P.T., "Public Philosophy: Distinction Without Authority?," *Journal of Medicine and Philosophy* 15:411-424, 1990.
55. Momeyer, R.W., "Philosophers and the Public Policy Process: Inside, Outside, or Nowhere at All?," *Journal of Medicine and Philosophy* 15:391-409, 1990.
56. Moskowitz, E.H., The Hastings Center, Briarcliff Manor, NY, remarks at "Biomedical Ethics in U.S. Public Policy," a workshop sponsored by the Office of Technology Assessment, U.S. Congress, Dec. 4, 1992.
57. Murray, T.H., "Speaking Unsmooth Things About the Human Genome Project," *Gene Mapping: Using Law and Ethics as Guides*, G.J. Annas and S. Elias (eds.) (New York, NY: Oxford University Press, 1992).
58. National Research Council, Committee for the Study of Inborn Errors of Metabolism, *Genetic Screening: Programs, Principles, and Research* (Washington, DC: National Academy Press, 1975).
59. Palca, J., "New Watchdogs in Washington," *Hastings Center Report* 23(2):5, 1993.
60. Povar, G.J., The George Washington University Medical Center, Washington, DC, remarks at "Biomedical Ethics in U.S. Public Policy," a workshop sponsored by the Office of Technology Assessment, U.S. Congress, Dec. 4, 1992.
61. Ramsey, P., *The Ethics of Fetal Research* (New Haven, CT: Yale University Press, 1975).
62. Rothman, D.J., *Strangers at the Bedside: A History of How Law and Bioethics Transformed Medical Decisionmaking* (New York, NY: Basic Books, Inc., 1991).
63. Ryan, K.J., Brigham and Women's Hospital, Boston, MA, personal communication, March 1993.
64. Sass, H.-M., "Blue-Ribbon Commissions and Political Ethics in the Federal Republic of Germany," *Journal of Medicine and Philosophy* 14:465-472, 1989.
65. Schrage, M., "'Bioethics' Burgeons, and Along With It Career Opportunities," *Washington Post*, Oct. 16, 1992.
66. Seldin, D.W., University of Texas Southwestern Medical School, Dallas, TX, remarks at "Biomedical Ethics in U.S. Public Policy," a workshop sponsored by the Office of Technology Assessment, U.S. Congress, Dec. 4, 1992.
67. Shapiro, D., Nuffield Council on Bioethics, London, United Kingdom, remarks at "Biomedical Ethics in U.S. Public Policy," a workshop sponsored by the Office of Technology Assessment, U.S. Congress, Dec. 4, 1992.
68. Stone, R., "Eyeing a Project's Ethics," *Science* 259:1820, 1993.
69. Stone, R., "NIH Adds and Extra Layer of Review for Sensitive Grants," *Science* 259:1820-1821, 1993.
70. Sun, M., "New Bioethics Panel Under Consideration," *Science* 222:34-35, 1983.
71. Tancredi, L. (ed.), Institute of Medicine, *Ethics in Health Care* (Washington, DC: National Academy Press, 1974).
72. U.S. Congress, House of Representatives, Committee on Government Operations, *Designing Genetic Information Policy: The Need for an Independent Policy Review of the Ethical, Legal, and Social Implications of the Human Genome Project*, H. Rpt. 102-478 (Washington, DC: U.S. Government Printing Office, 1992).
73. U.S. Congress, Office of Technology Assessment, *Human Gene Therapy—A Background Paper*, OTA-BP-BA-32 (Washington, DC: U.S. Government Printing Office, December 1984).
74. U.S. Congress, Office of Technology Assessment, *New Developments in Biotechnology: Ownership of Human Tissues and Cells*, OTA-BA-337 (Washington, DC: U.S. Government Printing Office, March 1987).
75. U.S. Congress, Office of Technology Assessment, *Life-Sustaining Technologies and the Elderly*, OTA-BA-306 (Washington, DC: U.S. Government Printing Office, July 1987).
76. U.S. Congress, Office of Technology Assessment, *Mapping Our Genes—The Human Genome Projects: How Big, How Fast?*, OTA-BA-373 (Washington, DC: U.S. Government Printing Office, April 1988).
77. U.S. Congress, Office of Technology Assessment, *Infertility: Medical and Social Choices*, OTA-BA-358 (Washington, DC: U.S. Government Printing Office, May 1988).
78. U.S. Congress, Office of Technology Assessment, *New Developments in Biotechnology: Patenting*

Life, OTA-BA-370 (Washington, DC: U.S. Government Printing Office, April 1989).
79. U.S. Congress, Office of Technology Assessment, *Neural Grafting: Repairing the Brain and Spinal Cord*, OTA-BA-462 (Washington, DC: U.S. Government Printing Office, September 1990).
80. U.S. Congress, Office of Technology Assessment, *Genetic Monitoring and Screening in the Workplace*, OTA-BA-455 (Washington, DC: U.S. Government Printing Office, October 1990).
81. U.S. Congress, Office of Technology Assessment, *Cystic Fibrosis and DNA Tests: Implications of Carrier Screening*, OTA-BA-532 (Washington, DC: U.S. Government Printing Office, August 1992).
82. U.S. Congress, Office of Technology Assessment, *The Biology of Mental Illness*, OTA-BA-538 (Washington, DC: U.S. Government Printing Office, September 1992).
83. U.S. Congress, Office of Technology Assessment, ''Biomedical Ethics in U.S. Public Policy,'' workshop transcript, Dec. 4, 1992.
84. U.S. Department of Health, Education, and Welfare, Ethics Advisory Board, *Report and Conclusions: Support of Research Involving Human In Vitro Fertilization and Embryo Transfer* (Washington, DC: U.S. Government Printing Office, 1979).
85. U.S. Department of Health, Education, and Welfare, National Commission for the Protection of Human Subjects of Biomedical and Behavioral Research, *Research on the Fetus: Report and Recommendations*, DHEW Pub. No. (OS)76-127 (Washington, DC: U.S. Government Printing Office, 1975).
86. U.S. Department of Health, Education, and Welfare, National Commission for the Protection of Human Subjects of Biomedical and Behavioral Research, *Research Involving Children* (Washington, DC: U.S. Government Printing Office, 1977).
87. U.S. Department of Health, Education, and Welfare, National Commission for the Protection of Human Subjects of Biomedical and Behavioral Research, *Research Involving Those Institutionalized as Mentally Infirm* (Washington, DC: U.S. Government Printing Office, 1978).
88. Walters, L., ''Commissions and Bioethics,'' *Journal of Medicine and Philosophy* 14:363-368, 1989.
89. Weisbard, A.J. and Arras, J.D., ''Commissioning Morality: An Introduction to the Symposium,'' *Cardozo Law Review* 6:223-241, 1984.
90. Williams, J.R., ''Commissions and Biomedical Ethics: The Canadian Experience,'' *Journal of Medicine and Philosophy* 14:425-444, 1989.
91. Yesley, M.S., ''The Use of an Advisory Commission,'' *Southern California Law Review* 51:1451-1469, 1978.
92. Yesley, M.S., Los Alamos National Laboratory, Los Alamos, NM, remarks at ''Biomedical Ethics in U.S. Public Policy,'' a workshop sponsored by the Office of Technology Assessment, U.S. Congress, Dec. 4, 1992.
93. Yesley, M.S., Los Alamos National Laboratory, Los Alamos, NM, personal communication, March 1993.

Appendix A International Bioethics Initiatives

In the past few years, bioethics has become a global enterprise, with commissions or institutes in the Americas, Africa, Asia, Australia, and Europe. Though no federally sponsored forum exists in the United States, other governments and multinational organizations increasingly are establishing working groups, committees, or commissions to deal with bioethical issues.

In November 1992, OTA conducted a mail survey of bioethicists and individuals in government offices in 47 countries. Individuals and groups in 35 countries generously responded to OTA's inquiry, and approximately 40 persons active in national and international bioethics committees abroad also contributed by telephone or in-person interviews. This appendix describes national bioethics commissions abroad—as well as a sampling of bioethics activity in regional and multinational groups—based on the survey and information collected through other reports and the literature. The appendix also notes some common themes about the structure and operations of international bioethics entities. Finally, although OTA attempted to be as comprehensive as possible, the limits inherent in a short-term survey are such that the absence of a description for any country does not mean a lack of interest or activity in that country.

COUNTRY-BY-COUNTRY ACTIVITIES

A broad range of bioethics initiatives occurs abroad. This section summarizes the wealth of information obtained on individual countries, with a particular focus on national bodies. (See also table 2-1.) The following section describes data on multinational efforts.

■ Argentina

In December 1992, the Secretary of Health, National Ministry of Health and Social Welfare, created the National Bioethics Commission. Its mandate includes establishment and oversight of research subject committees, assessment of research protocols, and consideration of other issues in bioethics. A National Commission of Bioethics and Health Research to enunciate policies and survey bioethics research programs has also been proposed (116).

■ Australia

A number of federal and state bioethics commissions have contributed to health policy development in Australia. For example, the New South Wales Law Reform Commission on Human Tissue Transplants influenced practice and policy in that field in the 1970s

(73). In 1982, the Medical Research Ethics Committee (MREC) was established within the National Health and Medical Research Council (NHMRC) (analogous to the U.S. National Institutes of Health) to formulate guidelines for medical research. Compliance with these guidelines was a condition of Federal funding of medical research (17,35).

A more general group, the National Bioethics Consultative Committee (NBCC), was established in 1988 by the Joint Meeting of federal and state Ministers of Health and Social Welfare. NBCC's role was to advise the Ministers, and it was composed of professionals from a number of fields. Its reports, primarily concerned with human reproduction, were intended to stimulate debate rather than reflect conventional, mainstream views. While the NBCC's work provoked much public discussion, it was opposed by influential groups including churches, feminist organizations, disability groups, right-to-life organizations, and most bioethics centers. The federal and state Ministers of Health and Social Welfare withdrew their support (47,119).

The functions of MREC and NBCC were consolidated in 1991 into the Australian Health Ethics Committee (AHEC) at the NHMRC (16,47,119). Like its predecessors, AHEC is multidisciplinary in membership, and its duties include policy development and public education, as well as monitoring institutional ethics committees. In 1992, the federal Parliament changed the status of AHEC to involve broader membership; it will also undertake wider community consultation and will no longer be subject to NHMRC in regard to issuing guidelines. AHEC is not viewed as representing the community, but rather challenging the community (47,119). In fact, it has lobbied against bioethical decisions made by state parliaments and committees (47,119).

While AHEC operates at a national level, the constitutional responsibility for regulating health care practice and research belongs to Australian states. Thus, it is at the state level that there has been the most bioethics activity, including exhaustive public consultation, consensus seeking, policy development, and legislation. Several committees of inquiry were established in states, and three states have on-going statutory committees composed of a minority of health professionals and a broad range of other professionals and lay people. Most such committees have examined single topics, such as genetic manipulation, informed consent, or reproductive technologies (17,47,66,119).

■ Brazil

Brazil's National Health Council considers bioethics in its deliberations. Established in 1937, but reorganized extensively in 1990, participants include government officials, health care providers, consumers, and researchers. Additionally, the Federal Council of Medicine, which was created in 1957 as part of the Professional Councils on Health, is a professional organization of doctors that operates independently of the government. It enforces an ethical code for doctors and oversees several regional councils that provide guidance to professionals (82,87).

■ Canada

Both the Canadian Government and several provinces have considered bioethical issues in the context of potential legislation recommended by law reform commissions. In particular, the work of the Canadian Federal Law Reform Commission, which concluded its work in 1992, contributed widely to discussions on bioethical issues. In addition, the Science Council of Canada, now disbanded, produced several bioethics studies (83).

In July 1984, the Medical Research Council (MRC) of Canada established a Standing Committee on Ethics in Experimentation which, with its Working Group on Research Involving Human Subjects, accepted outside consultation and recommended guidelines on research involving human subjects to the MRC. The MRC accepted the report as official MRC guidelines and required that all research it funds be in compliance with them (80). The National Council on Bioethics in Human Research was established in 1989 by the MRC, National Health and Welfare Canada, and the Royal College of Physicians and Surgeons. This body is an advisory group that aims to encourage ethical standards in biomedical and health research involving human subjects. The Council defines guidelines, advises institutional research ethics boards, and promotes professional and public education in research ethics (83).

Another effort is the legislatively established, single-issue Royal Commission on New Reproductive Technologies, which is responsible only to the privy council and the prime minister. It has been charged with performing a comprehensive and authoritative review of Canadian policies, attitudes, laws, and practices on current and forthcoming reproductive technologies. Like all Royal Commissions, it is well-funded; in fact, it is reportedly the most well-funded bioethics commission in the world—over CN$20 million, or nearly CN$1 per citizen. The Commission is chaired by a physician and has four other members and a large professional staff. Through more than 100 outside contracts, it has sponsored extensive research to review existing data. It also has held hearings and open meetings around Canada and expects to publish its reports as required in 1993 (69,110). Royal Commissions examining other bioethical issues can also be created on an ad hoc, time-limited basis (126).

■ China

China has no general bioethics commission, although the Ministry of Health created an Ad Hoc Expert Advisory Committee to draft a eugenics law. Membership consists of 20 scientists, physicians, and bioethicists. The Ministry is reportedly planning to create a standing committee on policy issues stemming from technological innovations in health care. The Chinese Academy of Social Sciences, an academic and policy institute that advises the government, also has an active program in bioethics. This program has sponsored several national conferences on bioethical issues, sometimes with support from the Ministry of Health (95).

■ Czech Republic

Czechoslovakia's 1989 revolution enabled the Czech Republic to create the Czech National Commission of Medical Ethics (Centralni Eticka Komise, or CEC) in 1990. CEC oversees the activities of numerous Local Commissions of Ethics, located in hospitals and other health care units that had previously existed in the former Czechoslovakia. CEC consists of 28 members appointed by the President of the Scientific Council of the Minister of Health Care and two staff. Members are not required to represent particular professions or ideologies and receive no payment. CEC meets once a month with no rules excluding public attendance (56).

■ Denmark

Currently, Denmark has two national bioethics commissions with overlapping areas of interest, and sometimes disagreement. The Central Scientific-Ethical Committee (CSEC) has been in operation since 1978 (58). It was created in the wake of the Helsinki II declaration on human subjects research and has been chaired by one of its drafters. It has a well-defined responsibility to oversee all clinical and laboratory medical science research involving humans (98). CSEC originally operated via a voluntary arrangement of professional groups and a government ministry, but in 1992 it was given statutory authority. Currently, it consists of two representatives from each regional human subjects research review board, a layperson, and a researcher. CSEC acts on disputed proposals and in cases where a matter of principle needs to be decided (15). A 1989 report on research involving human subjects by the Ministry of Health had a tangible impact on the current laws on the research ethical committee system (84,98). Also, in 1991, the Danish Medical Research Council gathered a working group to publish a report on scientific dishonesty (27).

In 1988, Parliament created the Danish Council of Ethics to consider a broader range of bioethical issues. The Council is the primary adviser to Parliament on ethical problems in the health sciences, excluding the research related questions handled by the CSEC (98). The Council appears to be more grassroots than any other in the world, making significant effort to mobilize public discussion. The Council's 17 members are predominately laypersons who are nearly evenly divided by gender (26). Though reports are written by Council members, the Council has a slightly larger staff—three professionals—than is common in Europe; the Council also includes academics for short periods of service (26).

The Council's public education efforts go far beyond anything attempted elsewhere. For example, in considering the definition of death and the ethical issues of protecting human gametes, the Council not only held public hearings, but financed local debates on television. It also distributed elaborate educational materials to high schools, produced award-winning films shown on national television and in movie theaters, produced booklets and brochures for public libraries, and sponsored creative contests for young people. One thousand articles have been published over 3 years, along with numerous editorials that drew hundreds of letters to the editor. Although public awareness of bioethics among the general population has been successfully raised—probably due to the small size and homogeneity of the population (108)—it is interesting to note that, even after widespread discussion among the public, the Council's own surveys revealed misunderstandings about brain death (26,39,58,92,98,100,101).

The Council's findings on the definition of death drew criticism from CSEC members. The two commissions also disagreed on the propriety of preserving brain tissue for research and teaching purposes; one of the provisions in the 1992 law establishing the CSEC's legal status was a directive to the two councils to cooperate.

■ Finland

The Finnish National Research Ethics Committee was established in 1991 as a permanent advisory body of the government. Its mandate is to make proposals and give expert statements to the government, function as an expert body, promote research ethics, participate in international research ethics cooperation, and inform the government about issues in research ethics. The Committee consists of 10 members representing the scientific fields and government authorities involved in research ethics. Members of the first Committee have been appointed until 1995, when new members will be appointed (70,77).

The Ethical Advisory Committee for the National Agency for Welfare and Health was founded in 1991 by the Ministry of Social Affairs and Health. Twenty members from various fields have been appointed to serve until 1994 by the Director General of the National Agency for Welfare and Health. The group promotes bioethics and provides advice about ethical issues. It has no formal power, but may propose regulations or legislation (65).

■ France

The French National Consultative Ethics Committee on Life and Medical Sciences (Comite Consultatif National d'Ethique Pour les Sciences de la Vie et de la Sante, or CCNE) was created in 1983 by the President (30). It is one of the first national bioethics committees abroad and was created to play a central role in the country's deliberations over bioethical issues (62). Additionally, France developed a system of local ethics committees in the 1980s. These committees focus mainly on human research, similar to the function of institutional review boards in the United States. The local ethics committees became subject to national regulation in 1990, but are completely independent of CCNE (89).

CCNE's 41 members are drawn from medicine, law, biology, nursing, social science, ethics, philosophy, and religion; 15 of 41 members are women (89). Members are not paid, although many devote major portions of their professional time to CCNE; staff size is small. The committee is housed at the Institut National de la Sante et de la Recherche Medicale (INSERM) (analogous to the National Institutes of Health) and meets in closed session. A 2-day public symposium is held each year to bring bioethical issues to the public's attention.

CCNE's mission, as defined by the decree establishing it, is to advise the government on questions of bioethics; questions can be brought to CCNE by members of the government, the presidents of the two houses of Parliament, or by public institutions involved in research. To date, CCNE has issued over 30 reports or statements on topics that include research on human subjects, embryo research, genetic testing, the use of fetal tissue for medical procedures, surrogate motherhood, testing of drug addicts in employment settings, use of RU486, and sex determination procedures at the Olympics.

CCNE's unusually large size permits wider representation of views and interests. It not only carries out studies of major bioethical issues, but also involves itself in day-to-day controversies arising in the hospitals and courts. High in visibility and prestige, its annual public meetings have been addressed on several occasions by the President of the Republic, and its deliberations and findings are covered extensively by the press.

Finally, the French Parliament currently is considering a wide ranging bioethics bill, which has cleared the Assembly and is awaiting Senate action (89,91). A National Center of Medical Ethics was created by presidential degree in 1992, but has not yet been assembled (12).

Germany

The former West Germany[1] has had a number of interministry ad hoc commissions and legislative ad hoc commissions. In 1985, a joint commission with representatives from several federal ministries and academic institutions issued a report on in vitro fertilization, genome analysis, and gene therapy. Known as the Benda Commission, this body was chaired by the President of the Constitutional Court. The report led the federal Ministry of Justice to introduce the Embryo Protection Law (45,103). The federal Parliament has also created ad hoc commissions, known as Enquete commissions, on bioethical issues. The Gentechnologie Enquete Commission focused on biotechnology and produced a report in 1987 (46). Another Enquete commission focused on technology assessment and the social acceptability of technology; a third, established in 1988, focused on the public health care system (102). Rather than simply fostering anticipatory moral debate, the Benda Commission and the Enquete commissions generally used worst case scenarios for determining the moral and social acceptability of modem medical technology, applied general principles such as "human dignity" to specific cases, and recommended the criminalization of future technological possibilities (103).

The Scientific Council of the Federal Chamber of Physicians established two ethics commissions in 1985 that involved philosophers, theologians, and health administrators; the Scientific Council examined issues of in vitro fertilization and embryo research (11). In 1986, the German Society for Medical Law addressed the issue of withholding treatment from severely handicapped newborns (33), and the German Society for Anthropology and Human Genetics issued guidelines concerning withholding information from parents about the sex of their fetus (32,103).

The former West Germany also has had ethics committees in medical schools, as well as chambers of physicians that assist physicians in consultation on and assessment of moral and legal issues of human experimentation. In 1986, an association was formed by these ethics committees to regulate their activities and memberships; it meets once a year (4).

The role of bioethics in German public policy is somewhat unusual in light of the Nazi era. In 1989, a visiting British bioethicist's lectures on bioengineering and mental retardation had to be canceled after anti-biotechnology and disability rights groups accused him of advocating the rekindling of the Nazi's "euthanasia" program of persons with disabilities. Subsequent to this, other public protests forced the cancellation of lectures and courses on bioethics. The anti-bioethics protesters feel that public discussion of contemporary bioethics will make "despicable and dangerous" views seem more respectable (125).

Greece

In 1987, a nongovernmental group called the Hellenic Society of Medical Ethics and Deontology was created by individuals interested in medical ethics. The Society then played a role in persuading the Ministry of Public Health to begin to establish a National Center of Medical Ethics and require an ethics committee in every Greek hospital (12,67,71). It also succeeded in obtaining legislation establishing a National Medical Ethics and Deontology Board, consisting of professionals in health, law, and theology, all to be appointed by the Minister of Health (67,71).

Hungary

The Hungarian Scientific Health Council established a Scientific and Research Ethics Committee in 1987. Comprised of 20 individuals, including physicians, theologians, ethicists, and lawyers, it is the parent forum overseeing human subjects research in Hungary. It coordinates the regional research ethics committees and defines and publishes unified principles for research ethics.

The Parliament Committee on Social, Health, and Family Welfare established a bioethics commission of 16 professionals, health officials, sociologists, and philosophers in 1990. It seeks to be an advisory forum for legislation and to advise lawmakers. Quarterly meetings are not formally announced and consequently are unknown to the public (10,63).

[1] Though OTA only has information on the former West Germany, this is not to imply that no bioethics activities took place in the former East Germany.

Israel

The Director General of the Israeli Ministry of Health convenes a Supreme Helsinki Committee—i.e., a human subjects review committee operating on provisions of the Helsinki Declaration, which has been incorporated into Israeli law—when research in sensitive areas is proposed. One such area has been "experiments regarding the human genetic code," and its focus in recent years has been reproductive medicine. This same subject will also be addressed by a new commission jointly created by the Ministries of Health and Justice. In addition, the Israeli Society for Medical Ethics has its own committee that has provided testimony to various public agencies (106).

Italy

The President of the Council of Ministers created a National Committee on Bioethics (Comitato Nazionale per le Bioetica) in 1990 to provide advice to the Parliament. Its functions include formulating opinions and proposing solutions to ethical and legal problems that could arise in conjunction with advances in research in the life and health sciences or in the development of clinical therapies. It is also mandated to promote the formation of codes of conduct for individuals in the life and health sciences and to promote the provision of accurate information to the public. By decree, topics include genetic therapy and safety of biological materials (127). Thirty-six members have been chosen based on disciplinary background, and four represent professional organizations, including the Italian Medical Association. The Committee meets in closed sessions and has no professional staff. The Committee strives for consensus, reserving the use of voting if necessary, and has produced more than 10 reports (107).

In 1990, the Prime Minister established a working group of scientists, physicians, philosophers, and legislators to produce legislation on the ethical and legal problems involved in assisted procreation procedures and in scientific research concerning human embryos (18). In addition, the Italian National Research Council has an 11-member Committee of Bioethics, as do several medical specialty societies, including the Italian Society of Neurology and the Italian Society of Fertility and Sterility. A number of local groups also exist (124).

Japan

Japan has no standing bioethics commission, but has established the Prime Minister's Ad Hoc Committee on Brain Death and Organ Transplantation—a subject that is highly controversial in Japan (59,78). Its deliberations were closed to the public. In January 1992, the Committee issued a report—with dissents—that endorsed a definition of brain death.

Additionally, Japan's Human Genome Project has granted funds for the study of ethical issues, and a 20-member Ethical, Legal, and Social Issues Working Group has been established. To date, it has convened two international conferences (50,114).

Luxembourg

In 1988, the government of Luxembourg established a National Ethical Consultative Commission for the Life and Health Sciences. The Commission is attached to the Ministry of the State and serves as an advisory agency to the government. It is responsible for the multidisciplinary study, either at its own initiative or at the request of the government, of the ethical aspects of various problems arising in the life and health sciences. It also examines the solutions and means to be employed to address these problems (96).

Malta

In 1989, Malta's Minister of Social Policy formed a National Ethical Health Committee (12). The government is also interested in developing policies and programs to enhance the lives of the elderly and to mitigate any negative effects resulting from the aging population's impact on development (3).

Mexico

In Mexico, a federally sponsored National Bioethics Commission (Comision Nacional de Bioetica) that reports to the Ministry of Health was established in 1992. The Commission's President is the Secretary of Health, who appoints the 10 commission members and an executive general secretary, all health professionals. The body holds monthly meetings and additional sessions when the need arises. Its goals include research, education of the public, and recommendations for legislation. Besides medical issues, its broad mandate includes oversight on environmental matters (85).

The Netherlands

The Netherlands has had a bioethics commission since 1977. The Health Council, the scientific advisory board of the government in the fields of public health and environmental hygiene, sponsors a permanent Commission on Health Ethics and Health Law (CHEHL). CHEHL is a standing advisory group that transmits findings to the government of subject-specific ad hoc committees organized by the Health Council. CHEHL's 10 to 20 members, all scientists, meet 6 times a year in closed sessions. Currently, the vice president of the Health Council chairs CHEHL. The staff consists of two lawyers and one ethicist. Among national bioethics commissions, only CHEHL is known to have undergone an evaluation, in this case from the Parliament in 1991; the outcome was favorable.

In 1989, the Minister of Health established the Dutch Interim Central Committee on Ethical Aspects of Medical Research (Kerncommissie Ethiek Medisch Onderzoek, or KEMO) and located it at the premises of the Health Council. KEMO is a national advisory commission for the assessment of planned medical research involving ethical, legal, and social issues. It directly advises local medical ethics boards of health care institutions, not the government; recommendations are nonbinding (18). From 1989 to 1990, KEMO met bimonthly and responded by confidential letter to requests for advice from four local ethics committees (72,97). In 1991, KEMO published its first annual report (1,36). The Federation of Health Care Organizations in the Netherlands and the Royal Dutch Medical Association have also been active in bioethics (29, 121,122).

New Zealand

New Zealand has a single governmental body responsible for bioethical issues, the Health Research Council Ethics Committee. Established in 1990, the Committee advises the Health Research Council on ethical issues related to research. It also advises ethics committees established by other bodies (e.g., hospitals) on standards, procedures, and membership. In New Zealand, the chair of an ethics committee and at least 50 percent of members are not health care professionals.

Other government activities include a review of assisted reproduction and reports commissioned through agencies such as the Medical Council and the Core Health Services Committee. Establishment of a National Health Ethics Committee is also under consideration (14). The Law Reform Division of the Department of Justice has also been active and, in March 1985, released a paper on artificial reproductive technologies (48).

Norway

In 1989, the Norwegian Parliament passed a law establishing three national research ethics committees. One, the National Committee for Medical Research Ethics already existed on a nonstatutory basis, and two were created to work in the fields of social sciences and science and technology (43). The National Committee for Medical Research Ethics has nine members, including two Members of Parliament. In the past, the Committee answered to the Medical Research Council, but it is now independent. Its professional staff consists of a theologian-bioethicist, a secretary, and a consultant. Additionally, a National Medical Ethics Committee has been proposed to investigate patient rights and health care rationing (75). Also in 1989, the Parliament voted 16 million kroner for the Medical Research Council to establish a Center for Biomedical Ethics in Oslo (43).

The Philippines

In 1987, the Philippine Council for Health Research and Development published *National Guidelines for Biomedical Research Involving Human Subjects* which, among other matters, created a National Ethics Committee and institutional review committees (93). These committees are primarily concerned with the ethical review of research activities, though they are expected eventually to review other medical and health care practices (28). The National Ethics Committee is dominated by nonphysicians: By law, it includes one homemaker, one attorney, one environmentalist, one social scientist, one representative of the religious community, one medical researcher, and one representative of the Philippine Medical Association. The membership of the National Ethics Committee is indicative of a trend in the Philippines toward greater public involvement in a previously physician-centered activity (28).

Poland

The Ethics Review Committee in Biomedical Research was founded in 1977 by the Ministry of Health and Social Welfare and has been active in bioethics. A Commission for Supervising Research on Human Subjects was also created in 1982 by the Ministry of Health and Social Welfare to advance proper policy for research involving human subjects; this commission is chaired by a physician in the Ministry of Health and Social Welfare, and the Minister appoints the members.

A Commission for Research Ethics was created in 1991 at the Scientific Council of the Ministry of Health and Social Welfare. This commission meets once a year in closed session to draft legal regulations for human subjects research, review and inspect research on human subjects, review local research ethics committees, and publish its findings (88,115).

Portugal

Following an initiative of the Parliament, the national government created the National Council on Ethics for the Life Sciences in 1990 (94). The group presents annual reports to the Prime Minister on the application of new technologies to human life and the relevant clinical and social implications. The 20 members serve 5-year terms, and membership consists of 10 medical professors or doctors of certain specialties, 4 jurists, 3 philosophers, 1 biologist, 1 engineer, and 1 Catholic moralist (18). The Council's chairperson is appointed by the Prime Minister; six members are elected by Parliament, and 14 others are appointed by ministries and by scientific and professional organizations. Monthly meetings are closed to the public, but press conferences are held regularly. The Council strives for consensus, but publishes dissents. A single administrative officer staffs the Council (105).

Romania

In 1990, the Academy of Medical Sciences established a Bioethics Committee out of a need to "correct the numerous deficiencies left by the 45 years of Communist regime" (79). It is independent of the government and consists of 20 members, including physicians, jurists, psychologists, and priests; the current president is a geneticist. Members meet four times a year and are not paid. The Committee has played a role in the withdrawal of a Ministry of Health draft law to legalize active euthanasia and in the establishment of bioethics groups in several clinics of pediatrics, surgery, and endocrinology (79). It has organized conferences on medical ethics and bioethics, and is currently attempting to develop ethics curricula in each Romanian medical school (12).

Russia

In 1992, a Russian National Committee on Bioethics was formed on the initiative of the Russian Academy of Sciences and the Russian Academy of Medical Sciences (102). Its tasks include identifying and defining ethical issues raised by recent advances in biomedical research and practice. The Committee's role is advisory, though it can make arrangements for examining and reporting on bioethical issues and formulating new guidelines or laws. The Committee recognizes the need to inform and promote public debate and discussion on bioethical issues, and it is expected to consult widely, publish reports, and make recommendations (102). The Committee's first act was organizing the Commission on Protection of Animals as Subjects of Scientific Experimentation (118). In 1991, the Committee cosponsored, with the United Nations Educational, Scientific, and Cultural Organization (UNESCO), an international symposium on bioethics (12).

Additionally, the Center of Biomedical Ethics and Law was organized in Moscow in 1990. The center conducts regular interdisciplinary debates on problems in Russian medicine, and will submit a proposal to the President of Russia to establish an all-Russian Committee on biomedical research. This committee would be involved in ethical assessment and regulation, and would not be controlled by medical or academic authorities (118).

South Africa

The South African Medical Research Council prepares ethical guidelines for medical research; a revised document will be issued in late 1993 (9). The South African Law Commission recently issued a draft report on surrogate motherhood (48).

Spain

The Ministry of Health has considered creating a national committee on bioethics, however, these discussions were recently suspended. The Ministry's multidisciplinary Advisory Council has examined several bioethical issues (99), and numerous government-sponsored ad hoc commissions also have studied bioethics (2). In 1990, legislation requiring that clinical trials and research projects involving drugs be approved by hospital ethics committees was enacted. Thus, the number of hospital ethics committees is rising in Spain. The Spanish Medical Association also has dealt with bioethics through its Commission of Deontology (99).

■ Sweden

Sweden's National Council on Medical Ethics has been functioning since 1985. It is advisory to the government, and works under the Ministry of Health and Social Affairs. The Council consists of 18 members, including 7 Members of Parliament and experts in ethics, the arts, and religion, as well as representatives of certain organizations. The Council's task is to shed light on fundamental medical ethics issues, keep abreast of state-of-the-art research, and to act as a link between science, the public, and political decisionmakers. The Council has the ability to choose topics on its own initiative. It meets in closed session, but holds a "day of ethics" to brings bioethics to the public's attention. The Council does not carry out its own investigations, but is designed to monitor trends and analyze problems in medical ethics.

Additionally, numerous ad hoc, topic-specific committees have also been formed by the Swedish government before it prepares a bill. These committees have published reports and influenced legislation, sometimes through the National Council on Medical Ethics. In 1987, the Minister of Health and Social Affairs appointed a special Swedish Committee on Transplantation, which has written a number of reports (111,112). Sweden's Medical Research Council houses a central committee that oversees local research ethics committees concerned with individual research projects (64,113).

■ Switzerland

Currently, no broad bioethics commission exists, but the Swiss Department of the Interior and Department of Justice and Police jointly created an Expert Commission on Human Genetics and Reproductive Medicine (the Amstad Commission), which met from 1986 to 1988. Its mandate was to discuss new reproductive technologies and their social, ethical, and legal impacts, point out abuses, and write recommendations for the government. The 21-member group met 17 times in closed session and released its report in 1989 (117). In addition, the Swiss Academy of Medical Sciences maintains a nongovernmental Central Ethics Commission that has issued guidelines to clinicians on a number of ethical matters (57).

■ Turkey

The Higher Council of Health, established by legislation in 1930, consists of nine health care professionals chosen by the Minister of Health. Its scope of activity includes some bioethics issues, but it does not publish its work. In the near future, a Central Ethics committee will be established in the Ministry of Health to administer human subjects guidelines through institutional review boards (90).

■ United Kingdom

In the United Kingdom, bioethics is incorporated into policy in several ways and in many institutions. The Medical Research Council publishes an ethics series that primarily focuses on human research subject issues. Most prominent among the bioethics councils was the 1982-84 Warnock Committee on embryos and reproductive technology under the auspices of the then Department of Health and Social Security (31); its recommendations were largely embodied in new legislation. Recently, the House of Lords established a Select Committee on Medical Ethics to consider a number of issues. The British Medical Association has a Medical Ethics Committee and the Royal Colleges (e.g., of Physicians or Psychiatrists) have also issued numerous guidelines and position papers. In addition, the General Medical Council has issued guidelines on the commerce of human organs (48).

Nevertheless, in contrast to other European nations, the government has rejected suggestions to create a national bioethics commission with a broader mandate (13); it has preferred to establish multiple committees that each offer advice on specific issues. Prominent bioethicists in the United Kingdom have complained that without a national commission, they cannot identify and pursue the "British position" on important issues in pan-European councils and conferences (123,124). Thus, with interest in a British commission rising, a private solution was pursued. Following extensive consultation with professional, scientific, legal, and consumer groups, the Nuffield Foundation, an educational and charitable trust, founded a private body in 1991. The Nuffield Council on Bioethics is designed to function similarly to governmental bioethics bodies elsewhere in Europe. Its 15 members, 8 of whom are women, do not represent constituencies, but were chosen with diversity in mind. The Council aims to stimulate coordination among the many parties now contributing to bioethics policy, to anticipate new problems, and to increase public awareness of the issues and their importance. Several working groups have been established. An executive secretary and two administrative assistants comprise the staff (107).

The Nuffield Council on Bioethics, as a private body, has no regulatory role; it is advisory only. Nevertheless, the Foundation's initiative was welcomed by the government. One staff member is government salaried and the Council is regarded as the national voice within the British bioethics community. In fact, in composition and procedures, the Council conducts business as if it had been created by the government. Nevertheless, to date, many European bioethics forums have been intergovernmental, and government officials represent the United Kingdom. Thus, whether the British government will accord the Nuffield Council the same influence and authority in intergovernmental deliberations over bioethical issues as other European counterparts enjoy is uncertain (53,54,68,76,107).

MULTINATIONAL ORGANIZATIONS

Beyond efforts tailored to individual countries, ongoing efforts exist to address multinational, cross-cultural, or shared concerns about bioethical issues. This section briefly reviews some of the activities sponsored by international groups.

■ United Nations System

In January 1993, the General Director of the United Nations Educational, Scientific, and Cultural Organization (UNESCO) asked its Scientific and Technical Group to submit proposals for establishing an International Consultative Committee of Bioethics (120). The Committee will consist of 40 experts; its first task will be considering whether an international convention on the human genome should convene. The convention would deal with subjects linked to bioethics and human rights, especially problems stemming from trade of human tissues and cells, the use of genetic tests, eugenics, and cloning (120). The United Nations Fund for Population Assistance (UNFPA) has also provided funds for conferences that included bioethical issues (104).

In March 1993, the United Nations Commission on Human Rights adopted a Decision on Human Rights and Bioethics that seeks to ensure that the life sciences develop in a manner respectful of human rights. The Commission also promotes exchanges between national consultative bodies (48).

WHO has developed *Guiding Principles on Human Organ Transplantation* (endorsed in 1991 by the World Health Assembly). WHO's Health Legislation Unit also serves as a global clearinghouse for legislation, codes, and other measures in the field of bioethics. WHO reports on legislation and associated literature in a quarterly journal (48).

The Pan American Health Organization (PAHO) serves as the regional office of WHO for the Americas. Through publications and consultations, PAHO has fostered the development of bioethics in Central and South America. PAHO published the first regional survey of Latin American bioethics (19). PAHO and the University of Chile are also currently planning the establishment of a Pan American Institute of Bioethics, to be located at the University of Chile in Santiago. The Institute, slated to begin its work in 1994, will provide a "permanent place for . . . discussion of bioethical subjects," and its primary mission will be to support research and training in bioethics for the region (49).

■ Council for International Organizations of Medical Sciences

The Council for International Organizations of Medical Sciences (CIOMS) is an international, nongovernmental organization established in 1949 by two United Nations agencies (WHO and UNESCO). It began as a vehicle to facilitate the exchange of scientific information in the medical sciences through coordination of international organizations of medical sciences and support of international congresses on medical sciences. Since 1966, CIOMS has focused less on purely scientific medical subjects and more on the social and cultural impacts of medical science. The main activity of CIOMS has become the convening of broadly based, multidisciplinary, and internationally representative conferences concerned with the impact of progress in biomedical science on society, and in fields such as bioethics, health policy, drug development, and medical education. CIOMS, with WHO, has published the proceedings of many conferences and offered international ethical guidelines on a number of topics (6,7,24,25,128). In 1985, CIOMS constituted a steering committee on bioethics with representatives from a range of professional backgrounds and geographical locales, which in turn has organized "international dialogues" on ethical issues arising from many subjects, including human genome research (5).

■ Council of Europe

The Council of Europe (CE), an intergovernmental organization that seeks consensus among its 26 members on cultural and human rights issues, is active in bioethics. Following a 1985 resolution presented by the French Minister of Justice to the European Ministerial Conference on Human Rights, CE created an Ad Hoc Committee of Experts on Bioethics (CAHBI) to further the interests of member states in bioethical issues (42). In 1989, it held its First Symposium on Bioethics (20,74). In 1992, CE elevated CAHBI to full legal status within CE and changed its name to the Steering Committee on Bioethics (Comite Directeur sur la Bioethique, or CDBI).

CDBI consists of a diverse group of professionals and civil servants from member states and has delivered a number of reports from its working groups to the CE Committee of Ministers. Its activities have included a recommendation on international exchange and transportation of human substances, an opinion requested by the Netherlands government on voluntary euthanasia, and a report on assisted human reproduction. CDBI also has been particularly interested in issues of genetic screening, genetic testing, as well as forensic applications of DNA tests (21,64).

CDBI aims "to fill the political and legal gaps that may result from the rapid development of biomedical sciences," but to do this it must achieve the consensus of member states. It deals with this challenge in several ways: by "promot[ing] constructive dialogue between the member states;" by attending to the "principles and values which must guide any regulation in bioethics;" and by making "special efforts in order to identify the fundamental points on which the member States are unanimous" (22).

In 1989, CAHBI considered a proposal to create a European Bioethics Committee, but in 1992 it judged this step premature (23). Nevertheless, the CE has held meetings of the chairpersons of various national ethics committees and special symposiums (64). CDBI is currently preparing, pursuant to a 1990 request from the Committee of Ministers, a Convention for Bioethics. The Convention will consist of a framework of fundamental principles, based loosely on the European Convention on Human Rights (86). The Convention will incorporate general principles rather than detailed regulations, though these statements of principles could eventually become the basis for detailed protocols (109). Currently, protocols in organ transplantation, medical research involving humans and embryos, and the use of genetic information for nonmedical purposes are under preparation (23). The Convention is expected to be ready in 1994 and will be open to nonmember states (64).

■ European Community

Bioethics has emerged as an important human rights element of the European Community's (EC) scientific research policy (38). The Commission of the European Community recently decided to undertake a number of initiatives in bioethics and has established several working groups. For example, the Working Group on Human Embryos and Research intends to "determin[e] the area of consensus . . . and development of a common [legal] code," with the goal of pressuring member states to enact legislation where regulations are weak or nonexistent (44). In 1992, the EC formed a working group on the ethics of biotechnology (44).

The Working Group on Ethical, Social, and Legal Aspects (ESLA) of Human Genome Analysis has educational functions, but has also been charged to "make recommendations for future Commission initiatives—including legislation." Its charter requires it to take account of specific documents on human rights, including the Universal Declaration of Human Rights issued by the United Nations in 1948. Following a call for proposals, ESLA has funded 18 studies on ethical issues. The studies vary from applied to theoretical and examine a variety of key issues in human genetics (38,41).

The EC also empaneled an international commission on ethical issues in reproductive technology, called the Working Party on Ethical and Legal Issues Raised by New Reproductive Technology, or the Glover Commission. Despite cultural differences among the members and very little staff, the Glover Commission produced a report that is distinctive in its dissection of the moral arguments and premises on which rival positions on these issues rely (124). Though CE had no plans for publication, the Commission's chair, on his own initiative, brought the Commission's findings to public attention by arranging for their publication by academic presses (54,55).

In 1991, the EC identified general biomedical ethics as a fundamental research area in the Biomedical and Health Specific Program (37,40). The EC has established a research program in bioethics that appears to be the world's sole general fund for investigator-initiated general bioethics research. Proposals were solicited, and grants were provided for work that evaluates biomedical ethics issues and assesses the social impact and risks of current biomedical and health research programs. To date, eight projects have been funded, including grants addressing organ transplantation, artificial procreation, and AIDS. Initial funding for the eight grants was approximately 1.9 million ECU. Total funding under the program, approved for 1990-94, is slated at 4.67 million ECU (8).

European Parliament

The European Parliament seeks technical advice from its Scientific and Technological Options Assessment (STOA) Programme. STOA recently commissioned a major study that provides an analysis of the status of bioethics in Europe (42). The preliminary report indicates that the goal of European bioethics is regulation to ensure safety and to protect fundamental human rights. The report finds that "generally the first step toward the creation of successful regulation is the constitution of ethics committees to study the consequences of the various biological and genetic technologies" (42). The report recognizes that not every country has established a national ethics committee, "so the theory of harmonization of regulation will be difficult to put into practice," though the report nevertheless urges inter-European harmonization of law and recommendations (42).

The STOA report also discuss bioethics in the CE and EC and finds that the scope of these efforts largely overlaps. According to one EC official, however, the CE and EC have different roles to play in bioethics (44). The former gives general recommendations; the latter can submit proposals on specific regulations to the European Parliament. The CE can inspire new law, but it will be the law of member states. The CE and the EC are now taking steps to ensure cooperative activity in the field (70).

Other International Organizations

Other multinational bioethics organizations also are being created and include academic, government-sponsored, and professional groups and societies.

With the support of two U.S. foundations—Ford and Rockefeller—and WHO and UNFPA, the International Islamic Center for Population Studies and Research, Al-Azhar University in Cairo, Egypt hosted the first conference on the ethics of human reproduction research in the Muslim world in December 1991. The conference resulted in the adoption of ethical guidelines for human reproduction research in the Muslim world and the creation of the first Ethics Committee for Human Reproduction Research at Al-Azhar University, Cairo, Egypt (60,61,104). Similarly, the Islamic Organization for Education, Sciences, and Culture collaborated with the Faculty of Sciences at Qatar University to organize a meeting in February 1993 on the ethical implications of and guidelines for genetic research (104).

In 1987, the Medical Research Council of Canada and Canada's Department of National Health and Welfare convened an International Summit Conference on Bioethics (81). In 1990, the International Association of Human Biologists and the Japan Society of Human Genetics convened an International Panel Discussion on Education and Ethics in Medical Genetics (51). Hosted by the National Health Council of the Netherlands, with the support of the EC, the International Association of Bioethics (IAB) held its inaugural congress in 1992. IAB, headquartered in Australia, provides a forum for diverse views on bioethical issues, but it does not take positions. The International Association of Law, Ethics, and Science gathers individuals interested in bioethics and publishes a journal partly devoted to the works of national and international bioethics commissions (113).

International bioethics has also received the support of the United States' Hastings Center, which has held international bioethics conferences in Eastern Europe and other areas (34). The Eubios Ethics Institute in Christchurch, New Zealand and Tsukuba, Japan is a nonprofit group that holds international bioethics conferences, and publishes proceedings and newsletters (52). Among other public and private international bioethics groups are: the Fundacion Dr. Jose Maria Mainetti Escuela Latinoamericana de Bioetica, the European Association of the Centres of Medical Ethics, the International Federation of Catholic Universities' International Study Group on Bioethics, the Nordic Council's Bioethics Group, and the European Society for Philosophy of Medicine and Health Care, which has a Section for Medical Ethics in the Nordic Countries (2,64,70,116).

COMMON THEMES

Existing international commissions vary, and it is impossible to reach conclusions linking structure to performance. Because the commissions of each country exist in a unique cultural, political, social, and moral climate, it is dangerous to generalize too broadly or to transfer specific details too directly. Nevertheless, OTA's survey revealed several common points among international commissions: scope, sponsorship, public access, professional dominance, evaluation and soundness, role, structure, and a national voice. In particular, many countries' activities in bioethics stem from their interest in human rights.

Until recently, most bioethics commission abroad have been topical—i.e., devoted to one or a small number of issues—and temporary. Topics were selected, in advance, by the sponsor. The French commission, however, is wide ranging and seemingly permanent, with the freedom to choose its own topics. Other commissions established in Europe since the founding of the French commission also have been general, self generating, and open ended. Among the most influential commissions have been some single-topic efforts, such as the Warnock Committee in the United Kingdom. The clear trend, however, is toward a permanent bioethics commission that addresses new issues as they arise (124).

The independence of the commission is regarded by all observers as essential to its authority. Whether based in the legislature or in the executive branch, all but the United Kingdom's commissions are public. Most answer to, and are located in, the ministries of health, in contrast to the United States' President's Commission, which was located administratively outside the departmental structure of the executive branch. Responses to OTA did not reveal that existing commissions were perceived as overly beholden to their ministries.

Most national commissions in other countries limit public access, and meetings are generally closed. In some cases, members of the public may offer their views through periodic public symposia. One reason offered for the lack of public access is that some commissions rule on particular cases requiring confidentiality.

All governments have tried to ensure membership of non-health care professionals. In some cases, physicians and scientists are a clear minority. No survey data exist regarding public perceptions of the commissions as independent versus captured by special interests, but where separate committees exist to oversee human subjects research, these tend to be perceived as protective of the interests of physicians and scientists—even when lay members are present or even a majority (124).

Though bioethics commissions can be evaluated for productivity, influence, and soundness, little has been done in any country to date. Fragmentary though the responses to OTA's questionnaire were, however, it was striking that respondents' academic credentials were inversely related to their opinions about the soundness of the bioethics commission reports (124). Complaints that findings of various commissions are poorly argued, or not argued at all, were common.

National bioethics commissions abroad differ in their basic purpose. In some instances, they are directly advisory to parliaments; their existence is justified by their government's perceived need to develop legislation on complex technological and scientific issues through a slower and more deliberative process than allowed by usual legislative procedures. Other commissions exist to stimulate and educate the public, and still others assume the role of distilling and articulating a national sensibility on bioethical matters.

Particularly in international councils, national bioethics committees are increasingly seen as defining their nation's position on bioethics issues. To this extent, committees, and sometimes their members, act as or are viewed as national spokespersons (124).

All foreign bioethics bodies have a chair and numerous members, though they vary in size by a factor of four. Larger bodies can be more representative, but sacrifice working efficiency (65). More striking is the difference in the size of the staff, and complaints about lack of staff were frequent among responses to OTA's survey. Most have few—one to two—though isolated initiatives have larger staffs and more senior individuals. Only Canada has provided its Royal commissions a staff comparable to that found in the United States.

APPENDIX A REFERENCES

1. Aartsen, J.G.M., Health Council of the Netherlands, The Hague, The Netherlands, personal communication, March 1993.
2. Abel, F., "Dynamics of the Bioethics Dialogue in a Spain in Transition," S.S. Connor and H.L. Fuenzalida-Puelma (eds.), *Bioethics: Issues and Perspectives* (Washington, DC: Pan American Health Organization, 1990).
3. Agius, E., "Caring for the Elderly and Malta's National Health Scheme," *Hastings Center Report* 19(4):S7-S8, 1989.
4. Arbeitskreis Medizinischer Ethikkommissionen, "Verfahrensgrundsaetze," E. Pueschel and H.M. Sass (eds.), *Der Hippokratische Eid in der Medizin und andere Dokumente Medizinischer Ethik* (Bochum, Federal Republic of Germany: Zentrum fuer Medizinische Ethik, 1986).
5. Bankowski, Z., Council for International Organizations of Medical Sciences, Geneva, Switzerland, personal communication, November 1992.
6. Bankowski, Z., Bryant, J.H., and Last, J.M. (eds.), *Ethics and Epidemiology: International Guidelines—Proceedings of the XXVth CIOMS Conference, Geneva, Switzerland, 7-9 November 1990,* (Geneva, Switzerland: Council of International Organizations of Medical Sciences, 1991).
7. Bankowski, Z. and Howard-Jones, N., *Human Experimentation and Medical Ethics—XVth CIOMS Round Table Conference, Manila, 13-16 September 1981*, (Geneva, Switzerland: Council of International Organizations of Medical Sciences, 1982).
8. Bardoux, C., Commission of the European Communities, Brussels, Belgium, personal communication, March 1993.
9. Benatar, S.R., University of Cape Town, Cape Town, South Africa, personal communication, March 1993.
10. Blasszauer, B., Medical University of Pecs, Pecs, Hungary, personal communication, November 1992.
11. Bundesaerztekammer, "Richtlinien zur Durchfuehrung von IVF und ET als Behandlungsmethode," *Deutsches Aerzteblatt* 82(22):1-8, 1985.
12. Byk, C., Paris, France, personal communication, March 1993.
13. Campbell, A.V., "Committees and Commissions in the United Kingdom," *Journal of Medicine and Philosophy* 14:385-401, 1989.
14. Campbell, A.V., University of Otago, Dunedin, New Zealand, personal communication, March 1993.
15. Central Scientific-Ethical Committee of Denmark, *Report for 1991* (Copenhagen, Denmark: Ministry of Education and Research, 1992).
16. Charlesworth, M., "Bioethics in Australia" *Bioethics Yearbook: Volume II Regional Developments in Bioethics 1989-1991* (Dordrecht, Netherlands: Klumer, 1992).
17. Charlesworth, M., North Carlton, Australia, personal communication, March 1993.
18. Commission of the European Community, Working Group on Human Embryos and Research, "European Survey of the State of Legislation on Human Embryo Research," February 1992.
19. Connor, S.S. and Fuenzalida-Puelma, H.L. (eds.), *Bioethics: Issues and Perspectives* (Washington, DC: Pan American Health Organization, 1990).
20. Council of Europe, *Europe and Bioethics: Proceedings of the First Symposium of the Council of Europe on Bioethics* (Strasbourg, France: Council of Europe, 1990).
21. Council of Europe, "CAHBI Final Activity Report (91)17, Addendum II Revised," Strasbourg, France, Dec. 17, 1991.
22. Council of Europe, "The Work of the Council of Europe in the Field of Bioethics," CAHBI/INF (92)1, Strasbourg, France, Mar. 6, 1992.

23. Council of Europe, "First Round Table of Ethics Committees: Rapporteur's Internal Document," Madrid, Spain, Mar. 24-25, 1992.
24. Council of International Organizations of Medical Sciences and World Health Organization, *International Guidelines for Ethical Review of Epidemiological Studies* (Geneva, Switzerland: Council of International Organizations of Medical Sciences, 1991).
25. Council of International Organizations of Medical Sciences and World Health Organization, *International Guidelines for Biomedical Research Involving Human Subjects* (Geneva, Switzerland: Council of International Organizations of Medical Sciences, 1993).
26. Danish Council of Ethics, *Fourth Annual Report* (Copenhagen, Denmark: Danish Council of Ethics, 1992).
27. Danish Medical Research Council, *Scientific Dishonesty and Good Scientific Practice* (Copenhagen, Denmark: Danish Medical Research Council, 1992).
28. de Castro, L.D., "The Philippines: A Public Awakening," *Hastings Center Report* 20(2):27-28, 1990.
29. de Wachter, M.A.M., "Euthanasia in the Netherlands," *Hastings Center Report* 22(2):23-30, 1992.
30. Decree No. 83-132 to establish a National Advisory Ethics Committee for the Life and Health Sciences, "Portant creation d'un Comite consultatif national d'etique pour les sciences de la vie et de la sante," *Journal Officiel de la Republique Francaise, Edition des Lois et Decrets*, Paris, France, Feb. 23, 1983.
31. Department of Health and Social Security, *Report of the Committee of Enquiry Into Human Fertilisation and Embryology, Cmnd. 9314* (London, United Kingdom: Her Majesty's Stationery Office, 1984).
32. Deutsche Gesellschaft fuer Anthropologie und Humangenetik, *Empfehlungen zur Chorionbiopsie*, manuscript, 1986.
33. Deutsche Gesellschaft fuer Medizinrecht, "Grenzen der aertlichen Behandlungspflicht bei schwerstgeschaedigten Neugeborenen. Empfehlungen," *Medizinrecht* 4:281-282, 1986.
34. Donnelley, S., "Hastings on the Adriatic," *Hastings Center Report* 20(6):5-6, 1990.
35. Dunne, R.M., Provincial Bioethics Center, South Brisbane, Australia, personal communication, November 1992.
36. Dutch Interim Central Committee on Ethical Aspects of Medical Research, *Annual Report 1989 & 1990* (The Hague, The Netherlands: Dutch Interim Central Committee on Ethical Aspects of Medical Research, 1991).
37. Elizalde, J, "Les Activities de la Communaute Europeenne dans le Domaine de l'Ethique Medicale," R. Dierkens (ed.), *Proceedings of the IXth World Congress of Medical Law v. 1* (Belgium: Rijksuniversiteit Gent, 1991).
38. Elizalde, J., "Bioethics as a New Human Rights Emphasis in European Research Policy," *Kennedy Institute of Ethics Journal* 2:159-170, 1992.
39. Eubios Ethics Institute, *Newsletter*, Mar. 3, 1993.
40. European Community, "Council Decision of 9 September 1991 Adopting a Specific Research and Technologies Development Programme in the Field of Biomedicine and Health," *Official Journal of the E.C.* L 267(24 April):25-32, 1991.
41. European Community, "Call for Proposals for Studies on the Ethical, Social and Legal Aspects of Human Genome Analysis," *Official Journal of the E.C.* C 212(14 August):15-16, 1991.
42. European Parliament STOA Programme, "Bioethics in Europe," preliminary contract report prepared by the Gruppo di Attenzione sulle Biotecnologie, Milan, Italy, January 1992.
43. "European Round-Up: Norway," *Bulletin of Medical Ethics*, No. 51, June 1989.
44. Fasella, P., Commission of the European Community, Brussels, Belgium, personal communications, December 1992, January 1993.
45. Federal Republic of Germany, Bundesminister fuer Justiz, *Entwurf eines Embryonenschutzgesetzes (E.Sch.G.)* (Bonn, Federal Republic of Germany: Bundesminister fuer Justiz, 1986).
46. Federal Republic of Germany, Enquetekommission des Deutschen Bundestages, *Bericht: Chancen und Risiken der Gentechnologie (Drucksache 10/6775)*, (Bonn, Federal Republic of Germany: Deutscher Bundestag, 1987).

47. Fleming, J.I., Southern Cross Bioethics Institute, Plympton, Australia, personal communication, December 1992.
48. Fluss, S.S. World Health Organization, Geneva, Switzerland, personal communication, March 1993.
49. Fuenzalida, H.L., Pan American Health Organization, Washington, DC, personal communication, January 1993.
50. Fujiki, N., Fukui Medical School, Matsuokacho Fukui Prefecture, Japan, personal communication, November 1992.
51. Fujiki, N., Bulqzhenkov, V., and Bankowski, Z. (eds.), *Medical Genetics and Society* (New York, New York: Kugler Publications, 1991).
52. Fujiki, N. and Macer, D.R.J., *Human Genome Research and Society: Proceedings of the Second International Bioethics Seminar in Fukui, 20-21 March, 1992* (Christchurch, New Zealand: Eubios Ethics Institute, 1992).
53. Gillon, R., Imperial College of Science, Technology and Medicine Health Service, London, United Kingdom, personal communication, November 1992.
54. Glover, J., Oxford University, Oxford, United Kingdom, personal communication, December 1992.
55. Glover, J., Oxford University, Oxford, United Kingdom, remarks at ''Biomedical Ethics in U.S. Public Policy,'' a workshop sponsored by the Office of Technology Assessment, U.S. Congress, Dec. 4, 1992.
56. Haderka, J., Palacky University, Havirov, Czech Republic, personal communication, November 1992.
57. Hitzig, W., Swiss Academy of Medical Sciences, Zurich, Switzerland, personal communication, November 1992.
58. Holm, S., University of Copenhagen, Copenhagen, Denmark, personal communication, November 1992.
59. Hoshino, K., Kyoto Women's University, Kyoto, Japan, personal communication, November 1992.
60. International Islamic Center for Population Studies and Research, *Ethical Guidelines for Human Reproduction Research in the Muslim World* (Cairo, Egypt: Al-Azhar University, Cairo, 1992).
61. International Islamic Center for Population Studies and Research, *Proceedings of the First International Conference on ''Bioethics in Human Reproduction Research in the Muslim World,'' 10-13 December 1991* (Cairo, Egypt: Al-Azhar University, 1992).
62. Isambert, F.-A., ''Ethics Committees in France,'' *Journal of Medicine and Philosophy* 14:445-456, 1989.
63. Jakab, T., Medical University of Pecs, Pecs, Hungary, personal communication, November 1992.
64. Jonsson, L., Ministry of Health and Social Affairs, Stockholm, Sweden, personal communications, November 1992, March 1993.
65. Karjalainen, S., National Agency for Welfare and Health, Helsinki, Finland, personal communication, November 1992.
66. Kasimba, P. and Singer, P., ''Australian Commissions and Committees on Issues in Bioethics,'' *Journal of Medicine and Philosophy* 14:403-424, 1989.
67. Katsas, A., Evangelismos Hospital, Athens, Greece, personal communication, November 1992.
68. Kennedy, I.M., King's College, University of London, London, United Kingdom, personal communication, November 1992.
69. Knoppers, B.M., University of Montreal Law Faculty, Montreal, Canada, remarks at ''Biomedical Ethics in U.S. Public Policy,'' a workshop sponsored by the Office of Technology Assessment, U.S. Congress, Dec. 4, 1992.
70. Kokkonen, P., National Agency for Welfare and Health, Helsinki, Finland, personal communications, November 1992, January 1993.
71. Kontogeorgas, K., Patras, Greece, personal communication, November 1992.
72. Kuitert, H., Free University, Amstelveen, The Netherlands, personal communication, January 1993.
73. Law Reform Commission, *Human Tissue Transplants, Report Number 7*, (Sydney, Australia: Commonwealth of Australia, 1977).
74. Le Bris, S., ''National Ethics Bodies,'' contract document for the Council of Europe, Ad Hoc Committee of Experts on Bioethics (CAHBI), Round Table of Ethics Committees, Madrid, Spain, Mar. 24, 1992.

75. Lie, R., University of Oslo, Oslo, Norway, personal communications, December 1992, January 1993.
76. Lockwood, M., University of Oxford, Oxford, United Kingdom, personal communication, November 1992.
77. Lopponen, P., Academy of Finland, Helsinki, Finland, personal communication, November 1992.
78. Macer, D., University of Tsukuba, Ibaraki, Japan, personal communication, November 1992.
79. Maximilian, C., Academy of Medical Sciences, Bucharest, Romania, personal communication, December 1992.
80. Medical Research Council of Canada, *Guidelines on Research Involving Human Subjects 1987* (Ottawa, Canada: Minister of Supply and Services Canada, 1987).
81. Medical Research Council of Canada, *Towards an International Ethic for Research with Human Beings: Proceedings of the International Summit Conference on Bioethics, April 5-10, 1987, Ottawa, Canada, Sponsored by the Medical Research Council of Canada and the Department of National Health and Welfare* (Ottawa, Canada: Minister of Supply and Services Canada, 1988).
82. Meira, A.R., University of Sao Paulo, Sao Paulo, Brazil, personal communication, March 1993.
83. Miller, J., National Council on Bioethics in Human Research, Ottawa, Canada, personal communication, March 1993.
84. Ministry of Health, *Research Involving Human Subjects: Ethics/Law* (Copenhagen, Denmark: Ministry of Health, 1989).
85. Moctezuma-Barragan, G., Direccion de Asuntos Juridicos de la Secretaria de Salud, Mexico, personal communication, December 1993.
86. Mundell, I., "Bioethics: Europe Drafts a Convention," *Nature* 356:368, 1992.
87. Neto, E.R., University of Brasilia, Brasilia, Brazil, personal communication, March 1993.
88. Nielubowicz, J., Ethics Review Committee in Biomedical Research, Warsaw, Poland, personal communication, December 1992.
89. Novaes, S.B., Centre National de la Recherche Scientifique, Paris, France, personal communication, March 1993.
90. Ors, Y., Ankara Medical Faculty, Ankara, Turkey, personal communication, November 1992.
91. Patel, T., "France Takes the Lead on Medical Ethics," *New Scientist* 136(1850):8, 1992.
92. Petersen, U.H., Ministry of Health, Copenhagen, Denmark, personal communication, November 1992.
93. Philippine Council for Health Research and Development, *National Guidelines for Biomedical Research Involving Human Subjects* (Manila, The Philippines: Philippine Council for Health Research and Development, 1987).
94. Portugal Law No. 14/90 of 9 June 1990 establishing the National Council on Ethics for the Life Sciences, *Diario de Republica*, Series I, 9 June 1990, No. 133.
95. Qiu, R.-Z., Chinese Academy of Social Sciences, Beijing, China, personal communications, December 1992, January 1993.
96. "Regulations of the Government in Council of 9 September 1988 Establishing a National Ethical Consultative Commission for the Life and Health Sciences," *Memorial: Journal Officiel du Grand-Duche de Luxembourg*, A-No 70, Dec. 27, 1988.
97. Rigter, H., Health Council of the Netherlands, The Hague, The Netherlands, personal communications, November 1992, January 1993.
98. Riis, P., Herlev University Hospital, Herlev, Denmark, personal communications, November 1992, March 1993.
99. Rivero, A.P., Universidad de Alcala, Madrid, Spain, personal communication, April 1993.
100. Rix, B.A., "The Importance of Knowledge and Trust in the Definition of Death," *Bioethics* 4:232-236, 1990.
101. Rix, B.A., The Danish Council of Ethics, Copenhagen, Denmark, personal communication, January 1993.
102. "Russian National Committee Set Up," *Bulletin of Medical Ethics*, October 1989.
103. Sass, H.M., "Blue-Ribbon Commissions and Political Ethics in the Federal Republic of Germany," *Journal of Medicine and Philosophy* 14:465-472, 1989.
104. Serour, G.I., Al-Azhar University, Cairo, Egypt, personal communications, December 1992, March 1993.

105. Serrao, D., University of Porto Medical School, Porto, Portugal, personal communications, December 1992, January 1993, March 1993.
106. Shapira, A., "Public Control of Biomedical Experiments," *The Use of Human Beings in Research*, A. de Vries and T. Engelhardt (eds.) (Dordrecht, The Netherlands: Kluwer Academic Publishers, 1988).
107. Shapiro, D., Nuffield Council on Bioethics, London, United Kingdom, personal communications, November 1992, March 1993.
108. Shapiro, D., Nuffield Council on Bioethics, London, United Kingdom, remarks at "Biomedical Ethics in U.S. Public Policy," a workshop sponsored by the Office of Technology Assessment, U.S. Congress, Dec. 4, 1992.
109. Simons, H., Department of Health, The Hague, The Netherlands, personal communication, January 1993.
110. Somerville, M.A., McGill Center for Medicine, Ethics, and Law, Montreal, Canada, personal communication, December 1992.
111. Swedish Committee on Transplantation, *Transplantation — Ethical, Medical and Legal Aspects* (Stockholm, Sweden: The Swedish Ministry of Health and Social Affairs, 1989).
112. Swedish Committee on Transplantation, *The Body After Death* (Stockholm, Sweden: The Swedish Ministry of Health and Social Affairs, 1992).
113. Swedish Ministry of Health and Social Affairs, *The National Council on Medical Ethics in Sweden* (Stockholm, Sweden: The Swedish Ministry of Health and Social Affairs, 1992).
114. Swinbanks, D., "Japan Bioethics: When Silence Isn't Golden," *Nature* 356:368, 1992.
115. Szawarski, Z., University College of Swansea, Swansea, Poland, personal communication, December 1992.
116. Tealdi, J.C., Fundacion Dr. José Maria Mainetti, Gonnet, Argentina, personal communication, December 1992.
117. Thevoz, J.-M., Foundation Louis Jeantet, Geneva, Switzerland, personal communication, November 1992.
118. Tichtchenko, P.D. and Yudin, B.G., "Toward a Bioethics in Post-Communist Russia," *Cambridge Quarterly of Healthcare Ethics* 4:295-303, 1992
119. Tonti-Filippini, N., Lower Templestone, Australia, personal communication, December 1992.
120. UNESCOPRESSE, News Service of the United Nations Educational, Scientific and Cultural Organization, press release 3(2), Jan. 22, 1993.
121. van Berkestijn, T.M.G., Borst-Eilers, E., Cohen, H.S., et al., "Meeting at Maastricht," *Hastings Center Report* 23(2):45, 1993.
122. van der Kloot Meijburg, H.H., Federation of Health Care Organizations in the Netherlands, Utrecht, The Netherlands, personal communication, November 1992.
123. Vines, G., "Research Ethics Face National Scrutiny," *New Scientist* 129(1758):14, 1991.
124. Wikler, D., "Bioethics Commissions Abroad," contract document prepared for the U.S. Congress, Office of Technology Assessment, January 1993.
125. Wikler, D. and Barondess, J., "Bioethics and Anti-Bioethics in Light of Nazi Medicine: What Must We Remember," *Kennedy Institute of Ethics Journal* 3:39-55, 1993.
126. Williams, J.R., "Commissions and Biomedical Ethics: The Canadian Experience," *Journal of Medicine and Philosophy* 14:425-444, 1989.
127. World Health Organization, "Italy," *International Digest of Health Legislation* 42(4):671-672, 1991.
128. World Health Organization and Council of International Organizations of Medical Sciences, *Proposed International Guidelines for Biomedical Research Involving Human Subjects* (Geneva, Switzerland: Council of International Organizations of Medical Sciences, 1982).

Appendix B

Legislation, Regulations, or Statutes for Previous U.S. Bioethics Initiatives

NATIONAL COMMISSION FOR THE PROTECTION OF HUMAN SUBJECTS OF BIOMEDICAL AND BEHAVIORAL RESEARCH

Reprinted below is Public Law 93-348, which established the National Commission for the Protection of Human Subjects of Biomedical and Behavioral Research. The National Commission operated from 1974-78.

Title II—Protection of Human Subjects of Biomedical and Behavioral Research

Part A—National Commission for the Protection of Human Subjects of Biomedical and Behavioral Research

Establishment of Commission

Section 201. (a) There is established a Commission to be known as the National Commission for the Protection of Human Subjects of Biomedical and Behavioral Research (hereinafter in this title referred to as the "Commission").

(b)(1) The Commission shall be composed of eleven members appointed by the Secretary of Health, Education, and Welfare (hereinafter in this title referred to as the "Secretary"). The Secretary shall select members of the Commission from individuals distinguished in the fields of medicine, law, ethics, theology, the biological, physical, behavioral and social sciences, philosophy, humanities, health administration, government, and public affairs; but five (and not more than five) of the members of the Commission shall be individuals who are or who have been engaged in biomedical or behavioral research involving human subjects. In appointing members of the Commission, the Secretary shall give consideration to recommendations from the National Academy of Sciences and other appropriate entities. Members of the Commission shall be appointed for the life of the Commission. The Secretary shall appoint the members of the Commission within sixty days of the date of the enactment of this Act.

(2)(A) Except as provided in subparagraph (B), members of the Commission shall each be entitled to receive the daily equivalent of the annual rate of the basic pay in effect for grade GS-18 if the General Schedule for each day (including traveltime) during which they are engaged in the actual performance of the duties of the Commission.

(B) Members of the Commission who are full-time officers or employees of the United States shall receive no additional pay on account of their service on the Commission.

(C) While away from their homes or regular places of business in the performance of duties of the Commission, members of the Commission shall be allowed travel expenses, including per diem in lieu of subsistence, in the same manner as persons employed intermittently in the Government service are allowed expenses under section 5703(b) of Title 5 of the United States Code.

(c) The chairman of the Commission shall be selected by the members of the Commission from among their number.

(d)(1) The Commission may appoint and fix the pay of such staff personnel as it deems desirable. Such personnel shall be appointed subject to the provisions of Title 5, United States Code, governing appointments in the competitive service, and shall be paid in accordance with the provisions of chapter 51 and subchapter III of chapter 53 of such title relating to classification and General Schedule pay rates.

(2) The Commission may procure temporary and intermittent services to the same extent as is authorized by section 3109(b) of Title 5 of the United States Code, but at rates for individuals not to exceed the daily equivalent of the annual rate of basic pay in effect for grade GS-18 of the General Schedule.

Commission Duties

Sec. 202. (a) The Commission shall carry out the following:

(1)(A) The Commission shall (i) conduct a comprehensive investigation and study to identify the basic ethical principles which should underlie the conduct of biomedical and behavioral research involving human subjects, (ii) develop guidelines which should be followed in such research to assure that it is conducted in accordance with such principles, and (iii) make recommendations to the Secretary (I) for such administrative action as may be appropriate to apply such guidelines to biomedical and behavioral research conducted or supported under programs administered by the Secretary, and (II) concerning any other matter pertaining to the protection of human subjects of biomedical and behavioral research.

(B) In carrying out subparagraphs (A), the Commission shall consider at least the following:

(i) The boundaries between biomedical or behavioral research involving human subjects and the accepted and routine practice of medicine.

(ii) The role of assessment of risk-benefit criteria in the determination of the appropriateness of research involving human subjects.

(iii) Appropriate guidelines for the selection of human subjects for participation in biomedical and behavioral research.

(iv) The nature and definition of informed consent in various research settings.

(v) Mechanisms for evaluating and monitoring the performance of Institutional Review Boards established in accordance with section 474 of the Public Health Service Act and appropriate enforcement mechanisms for carrying out their decisions.

(C) The Commission shall consider the appropriateness of applying the principles and guidelines identified and developed under subparagraph (A) to the delivery of health services to patients under programs conducted or supported by the Secretary.

(2) The Commission shall identify the requirements for informed consent to participation in biomedical and behavioral research by children, prisoners, and the institutionalized mentally infirm. The Commission shall investigate and study biomedical and behavioral research conducted or supported under programs administered by the Secretary and involving children, prisoners, and the institutionalized mentally infirm to determine the nature of the consent obtained from such persons or their legal representatives before such persons were involved in such research; the adequacy of the information given them respecting the nature and purpose of the research, procedures to be used, risks and discomforts, anticipated benefits from the research, and other matters necessary for informed consent; and the competence and the freedom of the persons to make a choice for or against involvement in such research. On the basis of such investigation and study the Commission shall make such recommendations to the Secretary as it determines appropriate to assure that biomedical and behavioral research conducted or supported under programs administered by him meets the requirements respecting informed consent identified by the Commission. For purposes of this paragraph, the term ''children'' means individuals who have not attained the

legal age of consent to participate in research as determined under the applicable law of the jurisdiction in which the research is to be conducted; the term ''prisoner'' means individuals involuntarily confined in correctional institutions or facilities (as defined in section 601 of the Omnibus Crime Control and Safe Streets Act of 1968 (42 U.S.C. 3781)); and the term ''institutionalized mentally infirm'' includes individuals who are mentally ill, mentally retarded, emotionally disturbed, psychotic, or senile, or who have other impairments of a similar nature and who reside as patients in an institution.

(3) The Commission shall conduct an investigation and study to determine the need for a mechanism to assure that human subjects in biomedical and behavioral research not subject to regulation by the Secretary are protected. If the Commission determines that such a mechanism is needed, it shall develop and recommend to the Congress such a mechanism. The Commission may contract for the design of such a mechanism to be included in such recommendations.

(b) The Commission shall conduct an investigation and study of the nature and extent of research involving living fetuses, the purposes for which such research has been undertaken, and alternative means for achieving such purposes. The Commission shall, not later than the expiration of the 4-month period beginning on the first day of the first month that follows the date on which all the members of the Commission have taken office, recommend to the Secretary policies defining the circumstances (if any) under which such research may be conducted or supported.

(c) The Commission shall conduct an investigation and study of the use of psychosurgery in the United States during the five-year period ending December 31, 1972. The Commission shall determine the appropriateness of its use, evaluate the need for it, and recommend to the Secretary policies defining the circumstances (if any) under which its use may be appropriate. For purposes of this paragraph, the term ''psychosurgery'' means brain surgery on (1) normal brain tissue of an individual, who does not suffer from any physical disease, for the purpose of changing or controlling the behavior or emotions of such individual, or (2) diseased brain tissue of an individual, if the sole object of the performance of such surgery is to control, change, or affect any behavioral or emotional disturbance of such individual. Such term does not include brain surgery designed to cure or ameliorate the effects of epilepsy and electric shock treatments.

(d) The Commission shall make recommendations to the Congress respecting the functions and authority of the National Advisory Council for the Protection of Subjects of Biomedical and Behavioral Research to be established under section 217(f) of the Public Health Service Act.

Special Study

Section 203. The Commission shall undertake a comprehensive study of the ethical, social, and legal implications of advances in biomedical and behavioral research and technology. Such study shall include—

(1) an analysis and evaluation of scientific and technological advances in past, present, and projected biomedical and behavioral research and services;

(2) an analysis and evaluation of the implementations of such advances, both for individuals and for society;

(3) an analysis and evaluation of laws and moral and ethical principles governing the use of technology in medical practice;

(4) an analysis and evaluation of public understanding of and attitudes toward such implications and laws and principles; and

(5) an analysis and evaluation of implications for public policy of such findings as are made by the Commission with respect to advances in biomedical and behavioral research and technology and public attitudes toward such advances.

Administrative Provisions

Section 204. (a) The Commission may for the purpose of carrying out its duties under sections 202 and 203 hold such hearings, sit and act at such times and places, take such testimony, and receive such evidence as the Commission deems advisable.

(b) The Commission may secure directly from any department or agency of the United States information necessary to enable it to carry out its duties. Upon the request of the chairman of the Commission, the head of such department or agency shall furnish such information to the Commission.

(c) The Commission shall not disclose any information reported to or otherwise obtained by it in carrying out its duties which (1) identifies any individual who has been the subject of an activity studied and investigated by the Commission, or (2) which concerns any information which contains or relates to a trade secret or other matter referred to in section 1905 of Title 18 of the United States Code.

(d) Except as provided in subsection (b) of section 202, the Commission shall complete its duties under sections 202 and 203 not later than the expiration of the 24-month period beginning on the first day of the first month that follows the date on which all the members of the Commission have taken office. The Commission shall make periodic reports to the President, the Congress, and the Secretary respecting its activities under sections 202 and 203 and shall, not later than ninety days after the expiration of such 24-month period, make a final report to the President, the Congress, and the Secretary respecting such activities and including its recommendations for administrative action and legislation.

(e) The Commission shall cease to exist thirty days following the submission of its final report pursuant to subsection (d).

Duties of the Secretary

Section 205. Within 60 days of the receipt of any recommendation made by the Commission under section 202, the Secretary shall publish it in the Federal Register and provide opportunity for interested persons to submit written data, views, and arguments with respect to such recommendation. The Secretary shall consider the Commission's recommendation and relevant matter submitted with respect to it and, within 180 days of the date of its publication in the Federal Register, the Secretary shall (1) determine whether the administrative action proposed by such recommendation is appropriate to assure the protection of human subjects of biomedical and behavioral research conducted or supported under programs administered by him, and (2) if he determines that such action is not so appropriate, publish in the Federal Register such determination together with an adequate statement of the reasons for his determination. If the Secretary determines that administrative action recommended by the Commission should be undertaken by him, he shall undertake such action as expeditiously as is feasible.

Part B—Miscellaneous

National Advisory Council for the Protection of Subjects of Biomedical and Behavioral Research

Section 211. (a) Section 217 of the Public Health Service Act is amended by adding at the end of the following new subsection:

''(f)(1) There shall be established a National Advisory Council for the Protection of Subjects of Biomedical and Behavioral Research (hereinafter in this subsection referred to as the ''Council'') which shall consist of the Secretary who shall be Chairman and not less than seven nor more than fifteen other members who shall be appointed by the Secretary without regard to the provisions of Title 5, United States Code, governing appointments in the competitive service. The Secretary shall select members of the Council from individuals distinguished in the fields of medicine, law, ethics, theology, the biological, physical, behavioral and social sciences, philosophy, humanities, health administration, government, and public affairs; but three (and not more than three) of the members of the Council shall be individuals who are or who have been engaged in biomedical or behavioral research involving human subjects. No individual who was appointed to be a member of the National Commission for the Protection of Human Subjects of Biomedical and Behavioral Research (established under Title II of the National Research Act) may be appointed to be a member of the Council. The appointed members of the Council shall have terms of office of four years, except that for the purposes of staggering the expiration of the terms of office of the Council members, the Secretary shall, at the time of appointment, designate a term of office of less than four years for members first appointed to the Council.

"(2) The Council shall—

"(A) advise, consult with, and make recommendations to, the Secretary concerning all matters pertaining to the protection of human subjects of biomedical and behavioral research;

"(B) review policies, regulations, and other requirements of the Secretary governing such research to determine the extent to which such policies, regulations, and requirements require and are effective in requiring observance in such research of the basic ethical principles which should underlie the conduct of such research and, to the extent such policies, regulations, or requirements do not require or are not effective in requiring observance of such principles, make recommendations to the Secretary respecting appropriate revision of such policies, regulations, or requirements; and

"(C) review periodically changes in the scope, purpose, and types of biomedical and behavioral research being conducted and the impact such changes have on the policies, regulations, and other requirements of the Secretary for the protection of human subjects of such research.

"(3) The Council may disseminate to the public such information, recommendations, and other matters relating to its functions as it deems appropriate.

"(4) Section 14 of the Federal Advisory Committee Act shall not apply with respect to the Council."

(b) The amendment made by subsection (a) shall take effect July 1, 1976.

Institutional Review Boards; Ethics Guidance Program

Section 212. (a) Part I of Title IV of the Public Health Service Act, as amended by section 103 of this Act, is amended by adding at the end the following new section:

"Institutional Review Boards; Ethics Guidance Program

"Section 474. (a) The Secretary shall by regulation require that each entity which applies for a grant or contract under this Act for any project or program which involves the conduct of biomedical or behavioral research involving human subjects submit in or with its application for such grant or contract assurances satisfactory to the Secretary that it has established (in accordance with regulations which the Secretary shall prescribe) a board (to be known as an 'Institutional Review Board') to review biomedical and behavioral research involving human subjects conducted at or sponsored by such entity in order to protect the rights of the human subjects of such research.

"(b) The Secretary shall establish a program within the Department under which requests for clarification and guidance with respect to ethical issues raised in connection with biomedical or behavioral research involving human subjects are responded to promptly and appropriately."

(b) The Secretary of Health, Education, and Welfare shall within 240 days of the date of the enactment of this Act promulgate such regulations as may be required to carry out section 474(a) of the Public Health Service Act. Such regulations shall apply with respect to applications for grants and contracts under such Act submitted after promulgation of such regulations.

Limitation on Research

Section 213. Until the Commission has made its recommendations to the Secretary pursuant to section 202(b), the Secretary may not conduct or support research in the United States or abroad on a living human fetus, before or after the inducted abortion of such fetus unless such research is done for the purpose of assuring the survival of such fetus.

Individual Rights

Section 214. (a) Subsection (c) of section 401 of the Health Programs Extension Act of 1973 is amended (1) by inserting "(1)" after "(c)", (2) by redesignating paragraphs (1) and (2) as subparagraphs (A) and (B), respectively, and (3) by adding at the end the following new paragraph:

"(2) No entity which receives after the date of enactment of this paragraph a grant or contract for biomedical or behavioral research under any program administered by the Secretary of Health, Education, and Welfare may—

''(A) discriminate in the employment, promotion, or termination of employment of any physician or other health care personnel, or

''(B) discriminate in the extension of staff or other privileges to any physician or other health care personnel, because he performed or assisted in the performance of any lawful health service or research activity, because he refused to perform or assist in the performance of any such service or activity on the grounds that his performance or assistance in the performance of such service or activity would be contrary to his religious beliefs or moral convictions, or because of his religious beliefs or moral convictions respecting any such service or activity.''

(b) Section 401 of such Act is amended by adding at the end the following new subsection:

''(d) No individual shall be required to perform or assist in the performance of any part of a health service program or research activity funded in whole or in part under a program administered by the Secretary of Health, Education, and Welfare if his performance or assistance in the performance of such part of such program or activity would be contrary to his religious beliefs or moral convictions.''

Special Project Grants and Contracts

Section 215. Section 772(a)(7) of the Public Health Service Act is amended by inserting immediately before the semicolon at the end thereof the following: '', or (C) providing increased emphasis on the ethical, social, legal, and moral implications of advances in biomedical research and technology with respect to the effects of such advances on individuals and society''.

Approved July 12, 1974.

ETHICS ADVISORY BOARD

Provisions governing the Ethics Advisory Board derive from volume 45, part 46, subpart B, of the Code of Federal Regulations (CFR) and the two charters under which it operated from 1978-80. The pertinent CFR sections and charters are reproduced below. (The regulations refer to ''Ethical Advisory Boards,'' but the body came to be known as the Ethics Advisory Board, as noted in the second charter.)

45 CFR, Subpart B—Additional Protections Pertaining to Research, Development, and Related Activities Involving Fetuses, Pregnant Women, and Human In Vitro Fertilization

Section 46.201 Applicability.

(a) The regulations in this subpart are applicable to all Department of Health and Human Services grants and contracts supporting research, development, and related activities involving: (1) The fetus, (2) pregnant women, and (3) human *in vitro* fertilization.

(b) Nothing in this subpart shall be construed as indicating that compliance with the procedures set forth herein will in an way render inapplicable pertinent State or local laws bearing upon activities covered by this subpart.

(c) The requirements of this subpart are in addition to those imposed under the other subparts of this part.

Section 46.202 Purpose.

It is the purpose of this subpart to provide additional safeguards in reviewing activities to which this subpart is applicable to assure that they conform to appropriate ethical standards and relate to important societal needs.

Section 46.203 Definitions.

As used in this subpart:

(a) ''Secretary'' means the Secretary of Health and Human Services and any other officer or employee of the Department of Health and Human Services to whom authority has been delegated.

(b) ''Pregnancy'' encompasses the period of time from confirmation of implantation (through any of the presumptive signs of pregnancy, such as missed menses, or by a medically acceptable pregnancy test), until expulsion or extraction of the fetus.

(c) ''Fetus'' means the product of conception from the time of implantation (as evidenced by any of the presumptive signs of pregnancy, such as missed menses, or a medically acceptable pregnancy test), until a determination is made, following expulsion or extraction of the fetus, that it is viable.

(d) ''Viable'' as it pertains to the fetus means being able, after either spontaneous or induced delivery, to survive (given the benefit of available medical therapy) to the point of independently maintaining heart beat and respiration. The Secretary may from time to time, taking into account medical advances, publish in the *Federal Register* guidelines to assist in determining whether a fetus is viable for purposes of this subpart. If a fetus is viable after delivery, it is a premature infant.

(e) ''Nonviable fetus'' means a fetus *ex utero* which, although living, is not viable.

(f) ''Dead fetus'' means a fetus *ex utero* which exhibits neither heartbeat, spontaneous respiratory activity, spontaneous movement of voluntary muscles, nor pulsation of the umbilical cord (if still attached).

(g) ''In vitro fertilization'' means any fertilization of human ova which occurs outside the body of a female, either through admixture of donor human sperm and ova or by any other means.

Section 46.204 Ethical Advisory Boards.

(a) One or more Ethical Advisory Boards shall be established by the Secretary. Members of these board(s) shall be so selected that the board(s) will be competent to deal with medical, legal, social, ethical, and related issues and may include, for example, research scientists, physicians, psychologists, sociologists, educators, lawyers, and ethicists, as well as representatives of the general public. No board member may be a regular, full-time employee of the Department of Health and Human Services.

(b) At the request of the Secretary, the Ethical Advisory Board shall render advice consistent with the policies and requirements of this part as to ethical issues, involving activities covered by this subpart, raised by individual applications or proposals. In addition, upon request by the Secretary, the Board shall render advice as to classes of applications or proposals and general policies, guidelines, and procedures.

(c) A Board may establish, with the approval of the Secretary, classes of applications or proposals which: (1) Must be submitted to the Board, or (2) need not be submitted to the Board. Where the Board so establishes a class of applications or proposals which must be submitted, no application or proposal within the class may be funded by the Department or any component thereof until the application or proposal has been reviewed by the Board and the Board has rendered advice as to its acceptability from an ethical standpoint.

(d) No application or proposal involving human *in vitro* fertilization may be funded by the Department of any component thereof until the application or proposal has been reviewed by the Ethical Advisory Board and the Board has rendered advice as to its acceptability from an ethical standpoint.

THE SECRETARY OF HEALTH, EDUCATION, AND WELFARE
WASHINGTON, D.C. 20201

CHARTER

ETHICAL ADVISORY BOARD

Purpose

In the Federal Register of August 8, 1975 (40 FR 33526), the Secretary of Health, Education, and Welfare published regulations regarding research, development, and related activities involving fetuses, pregnant women, and in vitro fertilization. The regulations, codified at 45 CFR Part 46, Subpart B, provide for the establishment by the Secretary of one or more Ethical Advisory Boards. Review of the current needs for such an advisory structure indicates that, for the time being, a single Board may be sufficient to meet the Department's needs. This Board will advise the Department regarding biomedical and behavioral research activities covered by Subpart B, in accordance with the provisions thereof. In addition, the Board will advise the Secretary, as requested, with respect to issues arising under Section 474(b) of the Public Health Service Act, as amended (42 U.S.C. 2891-3). The Board may also be assigned responsibility for advising with respect to ethical issues raised by other Departmental biomedical and behavioral research activities subject to the provisions of 45 CFR 46.

Authority

42 U.S. Code 217a. This Board is governed by the provisions of Public Law 92-463 which sets forth standards for the formation and use of advisory committees

Function

At the request of the Secretary or his designee, the Ethical Advisory Board shall render advice consistent with the policies and requirements of 45 CFR Part 46, Subpart B as to ethical issues, involving activities covered by that subpart, raised by individual applications or proposals. In addition, upon request by the Secretary or his designee, the Board shall render advice as to classes of applications or proposals and general policies, guidelines, and procedures under Subpart B.

The Board may establish, with the approval of the Secretary or his designee, classes of applications or proposals involving activities covered by Subpart B which: (1) must be submitted to the Board, or (2) need not be submitted to the Board. Where the Board so establishes a class of applcations or proposals which must be submitted, no application or proposal within the class may be funded by the Department of Health, Education, and Welfare or any component thereof until the application or proposal has been reviewed by the Board and the Board has rendered advice as to its acceptability from an ethical standpoint.

No application or proposal involving human <u>in vitro</u> fertilization may be funded by the Department of Health, Education, and Welfare or any component thereof until the application or proposal has been reviewed by the Ethical Advisory Board and the Board has rendered advice as to its acceptability from an ethical standpoint.

The Board must approve a request by the applicant or offeror to modify or waive requirements of Subpart B, in order for the Secretary or his designee to grant such a request.

At the request of the Secretary or his designee, the Board will provide advice with respect to issues arising under Section 474(b) of the Public Health Service Act, as amended (42 U.S.C. 2891-3).

The Secretary or his designee may assign the Board responsibility for advising the Department of Health, Education, and Welfare regarding ethical issues raised by other Department of Health, Education, and Welfare activities subject to the provisions of 45 CFR 46.

<u>Structure</u>

The Ethical Advisory Board shall consist of fourteen members, including the chairman, appointed by the Secretary or his designee. Members shall be so selected that the Board will be competent to deal with medical, legal, social, ethical, and related biomedical issues, provided that: (1) no more than seven may be scientists, of whom four shall be biomedical scientists and three social or behavioral scientists; and (2) the remainder shall be from other disciplines or representatives of the general public, except that at least one shall be an attorney and one shall be an ethicist. No Board member may be a regular, full-time employee of the Federal government.

Members shall be invited to serve for overlapping four-year terms, except that of these persons initially appointed to the Board, four shall be appointed for four-year terms, four for three-year terms, three for two-year terms, and three for one-year terms. Terms of more than two years are contingent upon renewal of the Board by appropriate action prior to its termination.

Management and staff services shall be provided by the Executive Secretary of the Ethical Advisory Board, Office of the Director, National Institutes of Health.

Meetings

Meetings will usually be held three times a year or at the call of the chairman, with the advance approval of a government official who shall also approve the agenda. A government official shall be present at all meetings.

Meetings shall be open to the public except as determined otherwise by the Secretary or his designee; notice of all meetings shall be given to the public.

Meetings shall be conducted, and records of proceedings kept, as required by applicable laws and Departmental regulations.

Decisions of the Board on matters of broad public interest shall be published in such form and manner as the Secretary may approve.

Compensation

Members shall be paid at the rate of $100 per day plus per diem and travel expenses in accordance with Standard Government Travel Regulations.

Annual Cost Estimate

Estimated annual cost for operating the Board, including compensation and travel expenses but excluding staff support, is $50,000. Estimate of annual person years of staff support required is two, at a cost of $40,000.

Report

An annual report shall be submitted to the Secretary and the Assistant Secretary for Health not later than 60 days following the beginning of the next fiscal year which shall contain, as a minimum, a list of members and their business addresses, the Committee's functions, the dates and places of meetings, and a summary of the Committee's activities and recommendations made during the fiscal year. A copy of the report shall be provided to the Department Committee Management Officer.

Termination Date

Unless renewed by appropriate action prior to its expiration, the Ethical Advisory Board will terminate two years from the date this charter is approved, or, if earlier, the first day of the first month that follows the date on which all the members of the National Advisory Council for the Protection of Subjects of Biomedical and Behavioral Research (as provided for by Section 211 of the National Research Act) are appointed.

APPROVED:

DEC 27 1976
Date

David Mathews
Secretary

THE SECRETARY OF HEALTH, EDUCATION, AND WELFARE
WASHINGTON, D. C. 20201

CHARTER

ETHICS ADVISORY BOARD

Purpose

The Ethics Advisory Board will review and advise the Secretary, as requested, with respect to the ethics of current and proposed departmental research and other activities, and of the missions, programs, agency assignments, or procedures that are proposed to, or are reviewed, supported, conducted, sponsored, monitored, or regulated by the Department or which may involve the Department directly or indirectly through other Federal, domestic, foreign, or international organizations, institutions, agencies or persons.

Department regulations (45 CFR 46) for the protection of human subjects involved in research, development or related activities conducted, supported, or sponsored by the Department provide for the establishment of one or more Ethics Advisory Boards within the Department. Review of the current needs for the prescribed functions indicates that, for the time being, a single Board may be sufficient to meet the Department's needs. This Board will advise the Department regarding research activities covered by applicable subparts and sections of 45 CFR 46, in accordance with the provisions thereof. In addition, the Board will advise the Secretary, as requested, with respect to issues arising under Section 474(b) of the Public Health Service Act, as amended (42 U.S.C. 289l-3).

The Board may initiate inquiries, hold hearings and conduct public meetings, symposia, or other means for the purpose of developing appropriate recommendations and to advise the Secretary.

Authority

42 U.S. Code 217a. This Board is governed by the provisions of Public Law 92-463 as amended which sets forth standards for the formation and use of advisory committees.

Functions

1. At the request of the Secretary, the Ethics Advisory Board shall advise, consult with, and make recommendations to the Secretary regarding the ethics of any research or other policy, or any mission, program, agency assignment, or activity.

2. The Board may conduct inquiries and hold hearings on proposed policies and regulations and on the interpretation, applicability, administration, and effectiveness of departmental regulations, policies or requirements and on the implementation of safeguards and assurances by institutions or by agencies within the Department for the purpose of protecting the rights and welfare of human subjects, or on other ethical matters and will report its findings and recommendations to the Secretary.

3. At the request of the Secretary, the Board will provide advice on ethical issues addressed to the Secretary regarding research, development or related activities involving human subjects.

4. The Board may advise, consult with and make recommendations to the Secretary on ethical issues that arise in regard to proposed or ongoing research, development or related activities involving human subjects that may be conducted, supported, sponsored or regulated by the Department.

5. The Board may consider appeals, requests and inquiries from Institutional Review Boards or comparable agency review committees that are addressed to the Secretary for guidance on ethical or policy questions regarding research activities or proposals involving human subjects when such referrals, in the judgment of the Secretary, present substantive ethical or related policy issues.

6. The Chairman of the Board shall report directly to the Secretary on all proposed agenda subjects, actions and recommendations of the Board on policy development or policy review matters.

Ethical reviews required by regulation or referred to or requested by the Board on specific proposals or ongoing activities will be reported through the Office of the Secretary to the head of the departmental agency or agencies involved.

Structure

The Ethics Advisory Board shall consist of no less than fourteen nor more than twenty members, including the Chairman, appointed by the Secretary.

Selection of members shall be made for representation from the legal, ethical, scientific, medical and social professions or from the general public, with special qualifications and competence to deal effectively with ethical issues of concern to the Department, provided that at least one member shall be an attorney, at least one member shall be an ethicist, at least one member shall be a practicing physician, at least one member shall be a theologian, and that no less than one-third nor more than half the total membership shall be career scientists with substantial research accomplishments, each to be selected for competency in one or more of the following categories; (a) basic biomedical and behavioral (e.g., physiological, genetic, psychological, pathological or etiological) research; (b) pediatrics, developmental human biology, obstetrical or gynecological research; (c) epidemiology, population or health services research; (d) psychiatric, clinical psychology, behavioral or sociological research; and (e) design and conduct of large scale clinical research programs for improving the treatment of major diseases or disorders.

No Board member may be a regular, full time employee of the Federal government.

Members shall be appointed for four year terms except that of the members first appointed, approximately one fourth shall be appointed for terms of one year, one fourth for terms of two years, one fourth for terms of three years, and one fourth for terms of four years.

Any member appointed to fill a vacancy occurring prior to expiration of the term for which his predecessor was appointed shall serve for the remainder of such term. Appointed members shall be eligible for reappointment for one additional four year term. Members may serve after the expiration of their terms until their successors have taken office.

A majority of the appointed members shall constitute a quorum of the Board.

The Board may refer specific questions, problems, and issues to any other Departmental committee, agency or staff for advice and counsel.

On recommendations by the Board, the Secretary may approve the issuance of requests for proposals and make sole or multiple contracts for personal or institutional services for consultants to the Board, to sponsor or conduct appropriate meetings, workshops, symposia, studies or investigations and for preparation of transcripts, reports, and other documents to assist the Board in its assigned functions.

The Board is empowered to have access to all records within the Department and to all records outside the Department available to the Secretary under the provisions of 45 CFR 46, or as may be additionally authorized by the Secretary or his designee.

Management and staff services shall be provided by the Office of the Director, National Institutes of Health. The Chairman of the Board, after consultation with the Board, shall appoint a Staff Director and a Deputy Staff Director. Additional staff shall be appointed by the Staff Director. The Office of the Secretary, HEW, will, as justified and necessary, on request of the Chairman and with concurrence by the Secretary, authorize detail assignments of staff from departmental agencies to the staff of the Board. The Staff Director may enter into contracts for the purpose of assisting the Board in the performance of its functions.

Meetings

The Board may hold up to ten meetings a year or at the call of the Chairman. Subcommittee meetings may also be held three to six times a year. All meetings require advance approval by an authorized government official who shall also approve the agenda. An authorized government official shall be present at all meetings.

Meetings shall be conducted, and records of proceedings kept, as required by applicable laws and departmental regulations.

Compensation

Members shall be paid at the rate of $182.72 per day plus per diem and travel expenses in accordance with Standard Government Travel Regulations.

Annual Cost Estimate

The estimated annual cost for the operations and functions of the Board including compensation, travel expenses and contract services but excluding staff support is estimated at $1,000,000. The estimated annual person years of staff support is ten, at a cost of approximately $270,000.

Report

An annual report shall be submitted to the Secretary no later than 60 days following the beginning of the next fiscal year which shall contain, as a minimum, a list of members and their business addresses, the Board's functions, the dates and places of meetings, and a summary of the Board's activities and recommendations made during the fiscal year. A copy of the report shall be provided to the Department Committee Management Officer.

Termination Date

Unless renewed by appropriate action prior to its expiration, the Ethics Advisory Board will terminate two years from the date this Charter is approved.

APPROVED:

JAN 11 1979
Date

Secretary

PRESIDENT'S COMMISSION FOR THE STUDY OF ETHICAL PROBLEMS IN MEDICINE AND BIOMEDICAL AND BEHAVIORAL RESEARCH

Public Law 95-622 authorized creation of the President's Commission for the Study of Ethical Problems in Medicine and Biomedical and Behavioral Research. Although not fully operational during the entire period, the authority for the President's Commission was 1978-83. Reprinted below is the relevant section of the U.S. Code.

42 U.S.C. Subchapter XVI—President's Commission for the Study of Ethical Problems in Medicine and Biomedical and Behavioral Research

Section 300 v. Commission

(a) Establishment; composition; appointment of members; vacancies

(1) There is established the President's Commission for the Study of Ethical Problems in Medicine and Biomedical and Behavioral Research (hereinafter in this subchapter referred to as the ''Commission') which shall be composed of eleven members appointed by the President. The members of the Commission shall be appointed as follows:

(A) Three of the members shall be appointed from individuals who are distinguished in biomedical or behavioral research.

(B) Three of the members shall be appointed from individuals who are distinguished in the practice of medicine or otherwise distinguished in the provision of health care.

(C) Five of the members shall be appointed from individuals who are distinguished in one or more of the fields of ethics, theology, law, the natural sciences (other than a biomedical or behavioral science), the social sciences, the humanities, health administration, government, and public affairs.

(2) No individual who is a full-time officer or employee of the United States may be appointed as a member of the Commission. The Secretary of Health and Human Services, the Secretary of Defense, the Director of Central Intelligence, the Director of the Office of Science and Technology Policy, the Administrator of Veteran's Affairs, and the Director of the National Science Foundation shall each designate an individual to provide liaison with the Commission.

(3) No individual may be appointed to serve as a member of the Commission if the individual has served for two terms of four years each as such a member.

(4) A vacancy in the Commission shall be filled in the manner in which the original appointment was made.

(b) Terms of members

(1) Except as provided in paragraphs (2) and (3), members shall be appointed for terms of four years.

(2) Of the members first appointed—

(A) four shall be appointed for terms of three years, and

(B) three shall be appointed for terms of two years, as designated by the President at the time of appointment.

(3) Any member appointed to fill a vacancy occurring before the expiration of the term for which his predecessor was appointed shall be appointed only for the remainder of such term. A member may serve after the expiration of his term until his successor has taken office.

(c) Chairman

The chairman of the Commission shall be appointed by the President, by and with the advice and consent of the Senate, from members of the Commission.

(d) Meetings

(1) Seven members of the Commission shall constitute a quorum for business, but a lesser number may conduct hearings.

(2) The Commission shall meet at the call of the Chairman or at the call of a majority of its members.

(e) Compensation; travel expenses, etc.

(1) Members of the Commission shall each be entitled to receive the daily equivalent of the annual rate of basic pay in effect for grade GS-18 of the General Schedule for each day (including travel time) during which they are engaged in the actual performance of duties vested in the Commission.

(2) While away from their homes or regular places of business in the performance of services for the Commission, members of the Commission shall be allowed travel expenses, including per diem in lieu of subsistence, in the same manner as persons employed intermittently in the Government service are allowed expenses under section 5703 of Title 5.

Section 300 v-1. Duties of Commission

(a) Studies and investigation; priority and order; report to President and Congress

(1) The Commission shall undertake studies of the ethical and legal implications of—

(A) the requirements for informed consent to participation in research projects and to otherwise undergo medical procedures;

(B) the matter of defining death, including the advisability of developing a uniform definition of death;

(C) voluntary testing, counseling, and information and education programs with respect to genetic diseases and conditions, taking into account the essential quality of all human beings, born and unborn;

(D) the differences in the availability of health services as determined by the income or residence of the persons receiving the services;

(E) current procedures and mechanisms designed (i) to safeguard the privacy of human subjects of behavioral and biomedical research, (ii) to ensure the confidentiality of individually identifiable patient records, and (iii) to ensure appropriate access of patients to information continued [sic] in such records, and

(F) such other matters relating to medicine or biomedical or behavioral research as the President may designate for study by the Commission.

The Commission shall determine the priority and order of the studies required under this paragraph.

(2) The Commission may undertake an investigation or study of any other appropriate matter which relates to medicine or biomedical or behavioral research (including the protection of human subjects of biomedical or behavioral research) and which is consistent with the purposes of this subchapter on its own initiative or at the request of the head of a Federal agency.

(3) In order to avoid duplication of effort, the Commission may, in lieu of, or as part of, any study or investigation required or otherwise conducted under this subsection, use a study or investigation conducted by another entity if the Commission sets forth its reasons for such use.

(4) Upon the completion of each investigation or study undertaken by the Commission under this subsection (including a study or investigation which merely uses another study or investigation), it shall report its findings (including any recommendations for legislation or administrative action) to the President and the Congress and to each Federal agency to which a recommendation in the report applies.

(b) Recommendations to agencies; subsequent administrative requirements

(1) Within 60 days of the date a Federal agency receives a recommendation from the Commission that the agency take any action with respect to its rules, policies, guidelines, or regulations, the agency shall publish such recommendation in the Federal Register and shall provide opportunity for interested persons to submit written data, views, and arguments with respect to adoption of the recommendation.

(2) Within the 180-day period beginning on the date of such publication, the agency shall determine whether the action proposed by such recommendation is appropriate, and, to the extent that it determines that—

(A) such action is not appropriate, the agency shall, within such time period, provide the Commission with, and publish in the Federal Register, a notice of such determination (including an adequate statement of the reasons for the determination), or

(B) such action is appropriate, the agency shall undertake such action as expeditiously as feasible and shall notify the Commission of the determination and the action undertaken.

(c) Report on protection of human subjects; scope; submission to President, etc.

The Commission shall biennially report to the President, the Congress, and appropriate Federal agencies on the protection of human subjects of biomedical and behavioral research. Each such report shall include a review of the adequacy and uniformity (1) of the rules, policies, guidelines, and regulations of all Federal agencies regarding the protection of human subjects of biomedical or behavioral research which such agencies conduct or support, and (2) of the implementation of such rules, policies, guidelines, and regulations by such agencies, and may include such recommendations for legislation and administrative action as the Commission deems appropriate.

(d) Annual report; scope; submission to President, etc.

Not later than December 15 of each year (beginning with 1979) the Commission shall report to the President, the Congress, and appropriate Federal agencies on the activities of the Commission during the fiscal year ending in such year. Each such report shall include a complete list of all recommendations described in subsection (b)(1) of this section made to Federal agencies by the Commission during the fiscal year and the actions taken, pursant [sic] to subsection (b)(2) of this section, by the agencies upon such recommendations, and may include such recommendations for legislation and administrative action as the Commission deems appropriate.

(e) Publication and dissemination of reports

The Commission may at any time publish and disseminate to the public reports respecting its activities.

(f) Definitions

For purposes of this section:

(1) The term ''Federal agency'' means an authority of the government of the United States, but does not include (A) the Congress, (B) the courts of the United States, and (C) the government of the Commonwealth of Puerto Rico, the government of the District of Columbia, or the government of any territory or possession of the United States.

(2) The term ''protection of human subjects'' includes the protection of the health, safety, and privacy of individuals.

Section 300 v-2. Administrative provisions

(a) Hearings

The Commission may for the purpose of carrying out this subchapter hold such hearings, sit and act at such times and places, take such testimony, and receive such evidence,as the Commission may deem advisable.

(b) Appointment and compensation of staff personnel; procurement and compensation of temporary and intermittent services; detail of personnel from other Federal agencies

(1) The Commission may appoint and fix the pay of such staff personnel as it deems desirable. Such personnel shall be appointed subject to the provisions of Title 5 governing appointments in the competitive service, and shall be paid in accordance with the provisions of chapter 51 and subchapter III of chapter 53 of such title relating to classification and General Schedule pay rates.

(2) The Commission may procure temporary and intermittent services to the same extent as is authorized by section 3109(b) of Title 5, but at rates for individuals not to exceed the daily equivalent of the annual rate of basic pay in effect for grade GS-18 of the General Schedule.

(3) Upon request of the Commission, the head of any Federal agency is authorized to detail, on a reimbursable basis, any of the personnel of such agency to the Commission to assist it in carrying out its duties under this subchapter.

(c) Contracting authority

The Commission, in performing its duties and functions under this subchapter, may enter into contracts with appropriate public or nonprofit private entities. The authority of the Commission to enter into such contracts is effective for any fiscal year only to such extent or in such amounts as are provided in advance in appropriation Acts.

(d) Information requirements and prohibitions

(1) The Commission may secure directly from any Federal agency information necessary to enable it to carry out this subchapter. Upon request of the Chairman of the Commission, the head of such agency shall furnish such information to the Commission.

(2) The Commission shall promptly arrange for such security clearances for its members and appropriate staff as are necessary to obtain access to classified information needed to carry out its duties under this subchapter.

(3) The Commission shall not disclose any information reported to or otherwise obtained by the Commission which is exempt from disclosure under subsection (a) of section 552 of Title 5 by reason of paragraphs (4) and (6) of subsection (b) of such section.

(e) Support services from Administrator of General Services

The Administrator of General Services shall provide to the Commission on a reimbursable basis such administrative support services as the Commission may request.

Section 300 v-3. Authorization of appropriations; termination of Commission

(a) To carry out this subchapter there are authorized to be appropriated $5,000,000 for the fiscal year ending September 30, 1979, $5,000,000 for the fiscal year ending September 30, 1980, $5,000,000 for the fiscal year ending September 30, 1981, and $5,000,000 for the fiscal year ending September 30, 1982.

(b) The Commission shall be subject to the Federal Advisory Committee Act, except that, under section 14(a)(1)(B) of such Act, the Commission shall terminate on December 31, 1982.

BIOMEDICAL ETHICS BOARD AND BIOMEDICAL ETHICS ADVISORY COMMITTEE

In 1985, Congress passed, and the President signed, Public Law 99-158, which included establishment of a congressional Biomedical Ethics Board and the appointment of the Biomedical Ethics Advisory Committee. Both entities ceased to exist in 1989, having been operational for approximately 16 months. Reprinted below is the relevant section of the U.S. Code.

42 U.S.C. Section 275. Biomedical Ethics Board

(a) Establishment

There is established in the legislative branch of the Government the Biomedical Ethics Board (hereinafter referred to as the "Board").

(b) Membership; term of office; vacancies; chairman and vice chairman; meetings

(1) The Board shall consist of twelve members as follows:

(A) Six Members of the Senate appointed as follows: Three members appointed by the Majority Leader of the Senate from the majority party and three members appointed by the Minority Leader from the minority party.

(B) Six Members of the House of Representatives appointed by the Speaker of the House of Representatives, three from the majority party and three from the minority party.

(2) The term of office of a member of the Board shall expire when the member leaves the office of Senator or Representative; as the case may be, or upon the expiration of eight years after the date of the member's appointment to the Board, whichever occurs first.

(3) Vacancies in the membership of the Board shall not affect the power of the remaining members to execute the functions of the Board and shall be filled in the same manner as in the case of the original appointment.

(4) The Board shall select a chairman and a vice chairman from among its members at the beginning of each Congress. The vice chairman shall act as chairman in the absence of the chairman or in the event of the incapacity of the chairman. The chairmanship and vice chairmanship shall alternate between the Senate and the House of Representatives with each Congress. The chairman during each even-numbered Congress shall be selected by the Members of the House of Representatives on the Board from among their number. The vice chairman during each Congress shall be chosen in the same manner from that House of Congress other than the House of Congress of which the chairman is a Member.

(5) The Board shall meet once every three months unless such meeting is dispensed with by the chairman, and may meet at any time upon the request of four or more members of the Board or upon the call of the chairman.

(c) Functions; annual report to Congress; report to the Congress on research and developments in genetic engineering

(1) The Board shall study and report to the Congress on a continuing basis on the ethical issues arising from the delivery of health care and biomedical and behavioral research, including the protection of human subjects of such research and developments in genetic engineering (including activities in recombinant DNA technology) which have implications for human genetic engineering.

(2)(A) Except as provided in subparagraph (B), an annual report shall be transmitted to the Congress identifying the issues which were the subject of the study conducted under paragraph (1) and identifying areas, programs, and practices of medicine and biomedical and behavioral research which have significant ethical implications and which would be appropriate subjects for study.

(B) A report on research and developments in genetic engineering (including activities in recombinant DNA technology) which have implications for human genetic engineering shall be transmitted to the Congress not later than eighteen months after the appointment of the Committee under subsection (d) of this section.

(d) Biomedical Ethics Advisory Committee; appointment, membership, compensation, etc.; functions; public hearings; availability of additional personnel and information; gifts and donations; use of mails

(1) To conduct the studies and make the reports required by subsection (c) of this section, the Board shall appoint a Biomedical Ethics Advisory Committee (hereinafter referred to as the "Committee"). The Committee shall consist of fourteen members as follows:

(A) Four of the members shall be appointed by the Board from individuals who are distinguished in biomedical or behavioral research,

(B) Three of the members shall be appointed by the Board from individuals who are distinguished in the practice of medicine or otherwise distinguished in the provision of health care.

(C) Five of the members shall be appointed by the Board from individuals who are distinguished in one or more of the fields of ethics, theology, law, the natural sciences (other than the biomedical or behavioral sciences), the social sciences, the humanities, health administration, government, and public affairs.

(D) Two of the members shall be appointed by the Board from individuals who are representatives of citizens with an interest in biomedical ethics but who possess no specific expertise.

(2)(A) The Committee, by majority vote, shall elect from its members a chairman and a vice chairman and appoint an executive director who shall serve for such time and under such conditions as the Committee may prescribe. In the absence of the chairman, or in the event of the incapacity of the chairman, the vice chairman shall act as chairman.

(B) The term of office of each member of the Committee shall be four years, except that any such member appointed to fill a vacancy occurring prior to the expiration of the term for which such member's predecessor was appointed shall be appointed for the remainder of such term. Terms of the members shall be staggered so as to establish a rotating membership.

(C) The members of the Committee shall receive no pay for their services as members of the Committee, but shall be allowed necessary travel expenses (or, in the alternative, mileage for use of privately owned vehicles and a per diem of subsistence at not to exceed the rate prescribed in sections 5702 and 5704 of Title 5) and other necessary expenses incurred by them in the performance of duties as a member of the Committee, without regard to the provisions of subchapter 1 [sic] of chapter 57 and section 5731 of Title 5, and regulations promulgated thereunder.

(D) The executive director of the Committee, with the approval of the Committee, may employ such staff and consultants as necessary to prepare studies and reports for the Committee.

(3)(A) The Committee may, for the purpose of carrying out its functions, hold such public hearings, sit and act at such times and places, and take such testimony, as the Committee considers appropriate.

(B) Upon request of the Committee, the head of any Federal agency is authorized to detail, on a reimbursable basis, any of the personnel of such agency to the Committee to assist the Committee in carrying out its functions.

(C) The Committee may secure directly from any department or agency of the United States information necessary to enable it to carry out its functions. Upon request of the chairman of the Committee, the head of such department or agency shall furnish such information to the Committee.

(D) The Committee may accept, use, and dispose of gifts or donations or services or property.

(E) The Committee may use the United States mails in the same manner and under the same conditions as other departments and agencies of the United States.

(e) Authorization of appropriations

To enable the Board and the Committee to carry out their functions there are authorized to be appropriated $2,000,000 for fiscal year 1986, $2,500,000 for fiscal year 1987, $3,000,000 for fiscal year 1988, $2,000,000 for fiscal year 1989, and $2,500,000 for fiscal year 1990.

Appendix C
Acknowledgments

OTA thanks the many individuals and organizations that generously supplied information for this study. In addition, OTA acknowledges the following individuals for their review of drafts of this background paper:

J.G.M Aartsen
Health Council of the Netherlands
The Hague, The Netherlands

Duane F. Alexander
National Institute of Child Health
and Human Development
Bethesda, MD

Margaret A. Anderson
American Public Health Association
Washington, DC

Paul W. Armstrong
Bridgewater, NJ

Christiane Bardoux
Commission of the European Communities
Brussels, Belgium

Nora K. Bell
University of South Carolina
Columbia, SC

S.R. Benatar
University of Cape Town
Cape Town, South Africa

Bela Blasszauer
Medical University of Pecs
Pecs, Hungary

Dan W. Brock
Brown University
Providence, RI

Christian Byk
Paris, France

Daniel Callahan
The Hastings Center
Briarcliff Manor, NY

Alastair V. Campbell
University of Otago
Dunedin, New Zealand

Arthur L. Caplan
University of Minnesota
Minneapolis, MN

Peter Carpenter
Stanford University
Atherton, CA

Max Charlesworth
North Carlton, Australia

R. Alta Charo
University of Wisconsin Law School
Madison, WI

James Childress
University of Virginia
Charlottesville, VA

Ellen Wright Clayton
The Vanderbilt Clinic
Nashville, TN

Robert M. Cook-Deegan
Institute of Medicine
Washington, DC

Jean Davies
World Federation of Right-to-Die Societies
Oxford, United Kingdom

Regis M. Dunne
Provincial Bioethics Center
South Brisbane, Australia

Gary B. Ellis
Office for Protection from Research Risks
Bethesda, MD

H. Tristram Englehardt, Jr.
Baylor College of Medicine
Houston, TX

Paolo Maria Fasella
Commission of the European Communities
Brussels, Belgium

John I. Fleming
Southern Cross Bioethics Institute
Plympton, Australia

John C. Fletcher
University of Virginia
Charlottesville, VA

S.S. Fluss
World Health Organization
Geneva, Switzerland

Patricia Anne Flynn
St. Mary's College of California
Moraga, CA

Hernan L. Fuenzalida
Pan American Health Organization
Washington, DC

K.W.M. Fulford
University of Oxford
Oxford, United Kingdom

Bradford H. Gray
Yale University
New Haven, CT

Jiri F. Haderka
Palacky University
Havirov, Czech Republic

D.A. Henderson
Office of Science and Technology Policy
Washington, DC

Angela R. Holder
Yale University
New Haven, CT

Søren Holm
University of Copenhagen
Copenhagen, Denmark

Kathy Hudson
Office of the Assistant Secretary for Health
Washington, DC

Ram Ishay
Israeli Society for Medical Ethics
Tel Aviv, Israel

Tibor Jakab
Medical University of Pecs
Pecs, Hungary

Albert R. Jonsen
University of Washington
Seattle, WA

Lena Jonsson
Ministry of Health and Social Affairs
Stockholm, Sweden

Eric T. Juengst
National Center for Human Genome Research
Bethesda, MD

Aristoteles G. Katsas
Evangelismos Hospital
Athens, Greece

George Kisch
Jestetten, Germany

Bartha Maria Knoppers
University of Montreal Law Faculty
Montreal, Canada

Walter Landesman
Sindicato Medico del Uruguay
Montevideo, Uruguay

Nicole Lery
Laboratoire de Medecine Legale et de Toxicologie Medicale
Lyon, France

Robert J. Levine
Yale University School of Medicine
New Haven, CT

Roberto Llanos-Zuloaga
Clínica Ricardo Palma
Lima, Peru

Fernando Lolas
University of Chile
Santiago, Chile

Charles R. McCarthy
Kennedy Institute of Ethics
Washington, DC

Affonso Renato Meira
University of Sao Paulo
Sao Paulo, Brazil

Judith Miller
National Council on Bioethics in Human Research
Ottawa, Canada

Gonzalo Moctezuma Barragán
Dirección de Asuntos Juridicos
Mexico City, Mexico

Maurizio Mori
Politeia
Milan, Italy

Ellen H. Moskowitz
The Hastings Center
Briarcliff Manor, NY

Arno G. Motulsky
University of Washington
Seattle, WA

Thomas H. Murray
Case Western Reserve University
Cleveland, OH

Simone B. Novaes
Centre National de La Recherche Scientifique
Paris, France

Robert S. Olick
Lowenstein, Sandler, Kohl, Fisher & Boylan
Roseland, NJ

Yaman Örs
Ankara Medical Faculty
Ankara, Turkey

Marian Osterweis
Association of Academic Health Centers
Washington, DC

Joan P. Porter
Office for Protection from Research Risks
Bethesda, MD

Gail J. Povar
George Washington University Medical Center
Washington, DC

Philip R. Reilly
Shriver Center for Mental Retardation, Inc.
Waltham, MA

Povl Riis
Herlev University Hospital
Herlev, Denmark

Eleutério Rodriguez Neto
University of Brasilia
Brasilia, Brazil

Maria Teresa Rotondo
Sindicato Medico del Uruguay
Montevideo, Uruguay

Kenneth J. Ryan
Brigham and Women's Hospital
Boston, MA

Gamal I. Serour
Al-Azhar University
Cairo, Egypt

Daniel Serrao
University of Porto Medical School
Porto, Portugal

Amos Shapira
Tel Aviv University
Ramat Aviv, Israel

David Shapiro
Nuffield Council on Bioethics
London, United Kingdom

Judith Swazey
The Acadia Institute
Bar Harbor, ME

Juan Carlos Téaldi
Fundación Dr. José Maria Mainetti
Gonnet, Argentina

Jean-Marie Thevoz
Fondation Louis Jeantet
Geneva, Switzerland

Nicholas Tonti-Filippini
Lower Templestowe, Australia

Harold Y. Vanderpool
University of Texas Medical Branch
Galveston, TX

LeRoy B. Walters
Kennedy Institute of Ethics
Washington, DC

Daniel J. Wikler
University of Wisconsin
Madison, WI

Susan M. Wolf
Harvard University
Cambridge, MA

Michael S. Yesley
Los Alamos National Laboratory
Los Alamos, NM

Index